AMERICAN INFRARED SURVEY

A Celebration of Infrared Photography

Edited By
Stephen Paternite and David Paternite

Photo Survey Press Publishing
Akron, Ohio

During the latter part of 1981, the possibilities of an American infrared survey were discussed. Our own involvement with the creative aspects of infrared photography went back many years and we felt there was an ever increasing number of photographers across the country who were either experimenting with or using infrared film exclusively as a form of visual communication. The first months of 1982 were devoted to organizing the survey guidelines which were sent to schools, workshops, galleries and individuals known to use infrared film. National ads were placed in the May and June issues of five major photo magazines. By the end of July, we had received over 1200 images from 330 photographers. After months of careful consideration, 80 images were selected that we felt best illustrated the vast creative potential and magical qualities of infrared photography.

Acknowledgements

We would like to express our appreciation to the following people for their encouragement and support: Mr. & Mrs. Edward J. Bargetz, Ms. Diane M. Hardy, Dr. & Mrs. A. H. Kyriakides, Mr. & Mrs. Joseph F. LaRose and Dr. & Mrs. Carl J. Paternite.

We would also like to extend our gratitude to the following individuals for their many suggestions and technical assistance: Robert B. Arnold, Tom Byers, Ron France, Gene Gambill, Joe LaRose, Rob Mooney, Drew Peretzky, Kathy Rineer and Jill Sands.

A very special thanks to Jim Peter, without whose help this book might not have seen publication.

We are grateful to Dave Gardner and his staff at Gardner/Fulmer Lithograph for their fine laser-scanned reproductions.

And finally, our warmest thanks to all the participating photographers for their sensitive, inventive images and enthusiasm for this project.

S.P. & D.P.

Copyright © 1982 by Photo Survey Press Publishing. All rights reserved in all countries. No part of this publication may be reproduced, translated or transmitted in any form or by any means without the prior written permission of Photo Survey Press Publishing, P.O. Box 9157, Akron, Ohio 44305.

Library of Congress Catalogue Card Number: 82-61960
ISBN: 0-9609812-0-9 $21.95

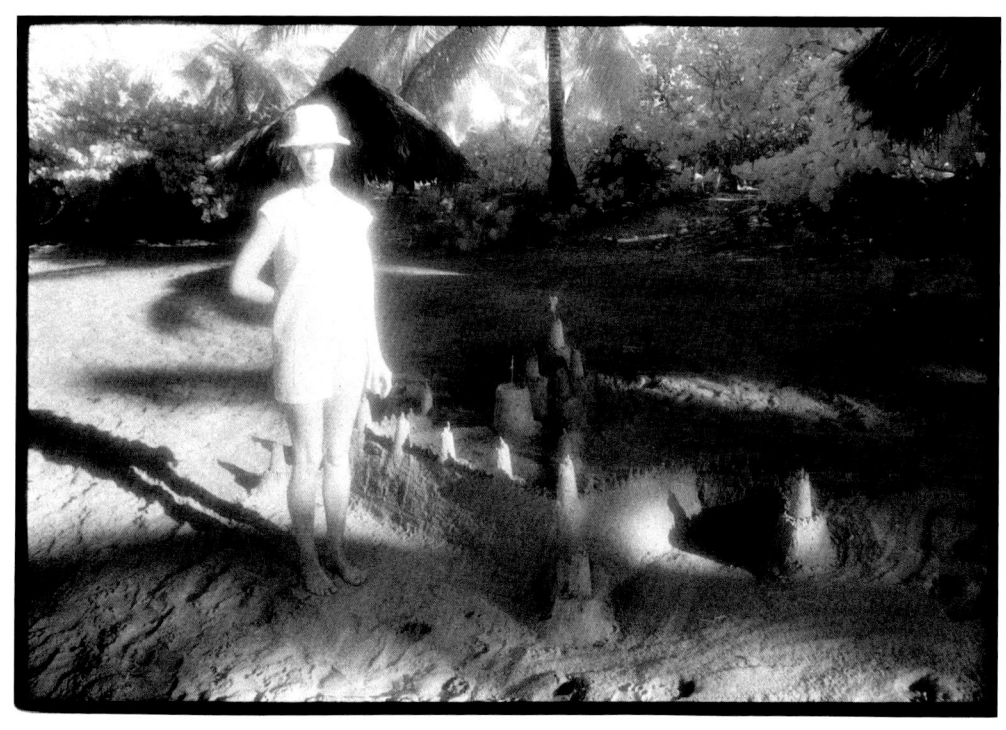

MARK P. TAYLOR
Jensen Beach, Florida
Lady With Sand Castle

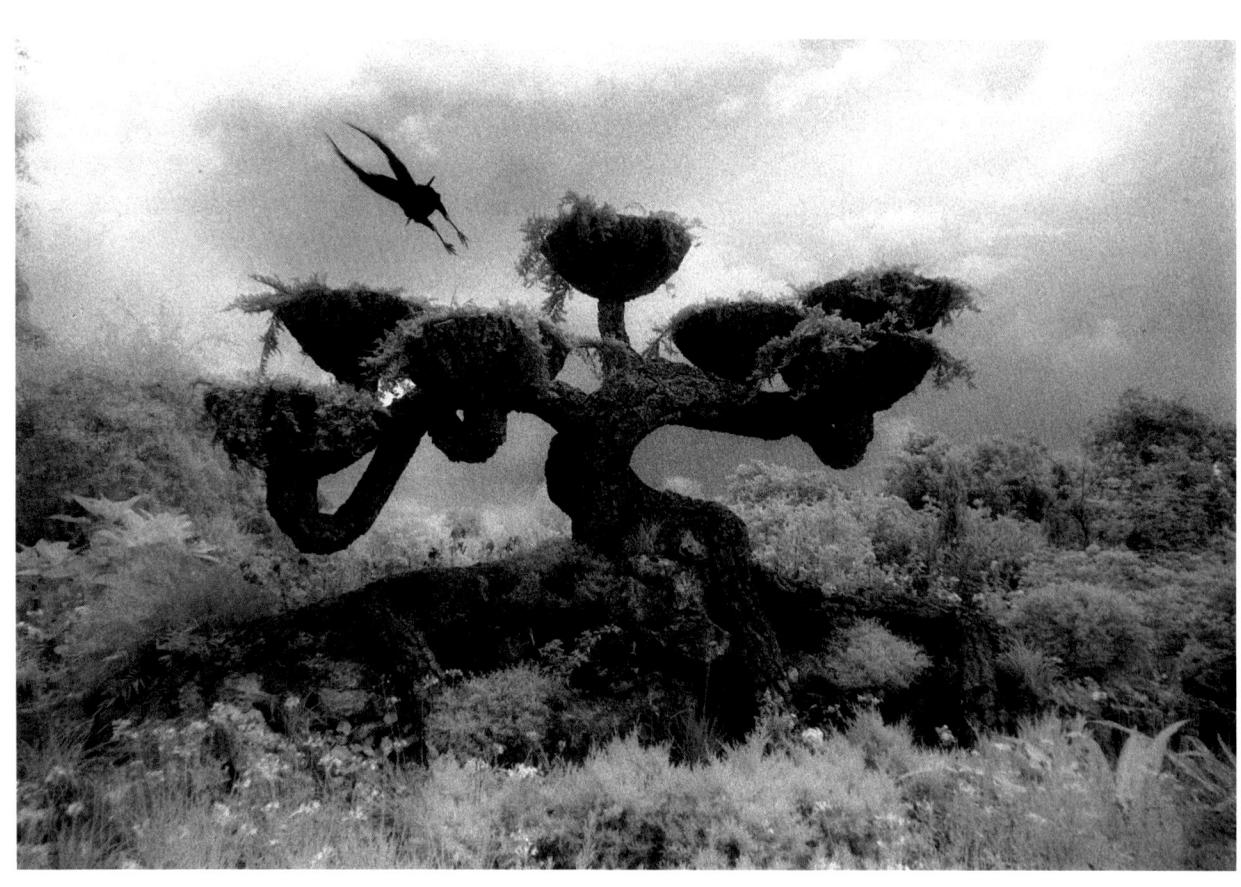

MERRY MOOR WINNETT
Greensboro, North Carolina

Shangri-La Tree

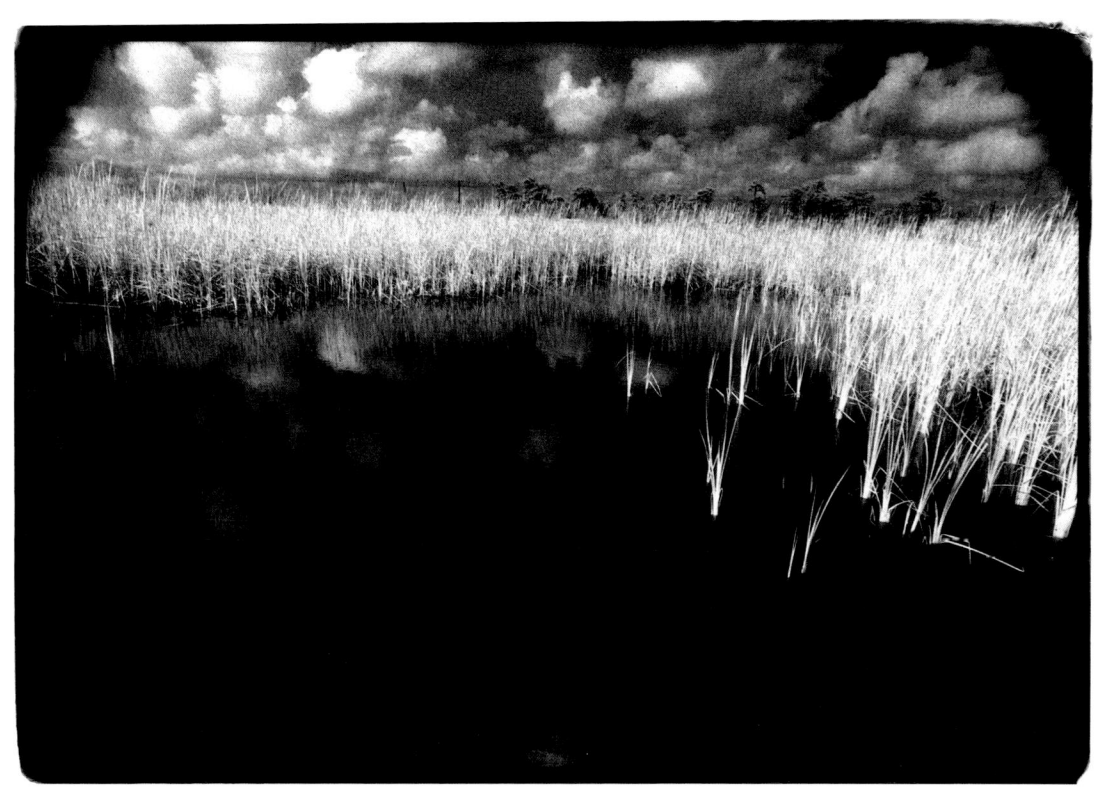

MERRY MOOR WINNETT
Greensboro, North Carolina
Pondscape, 1978

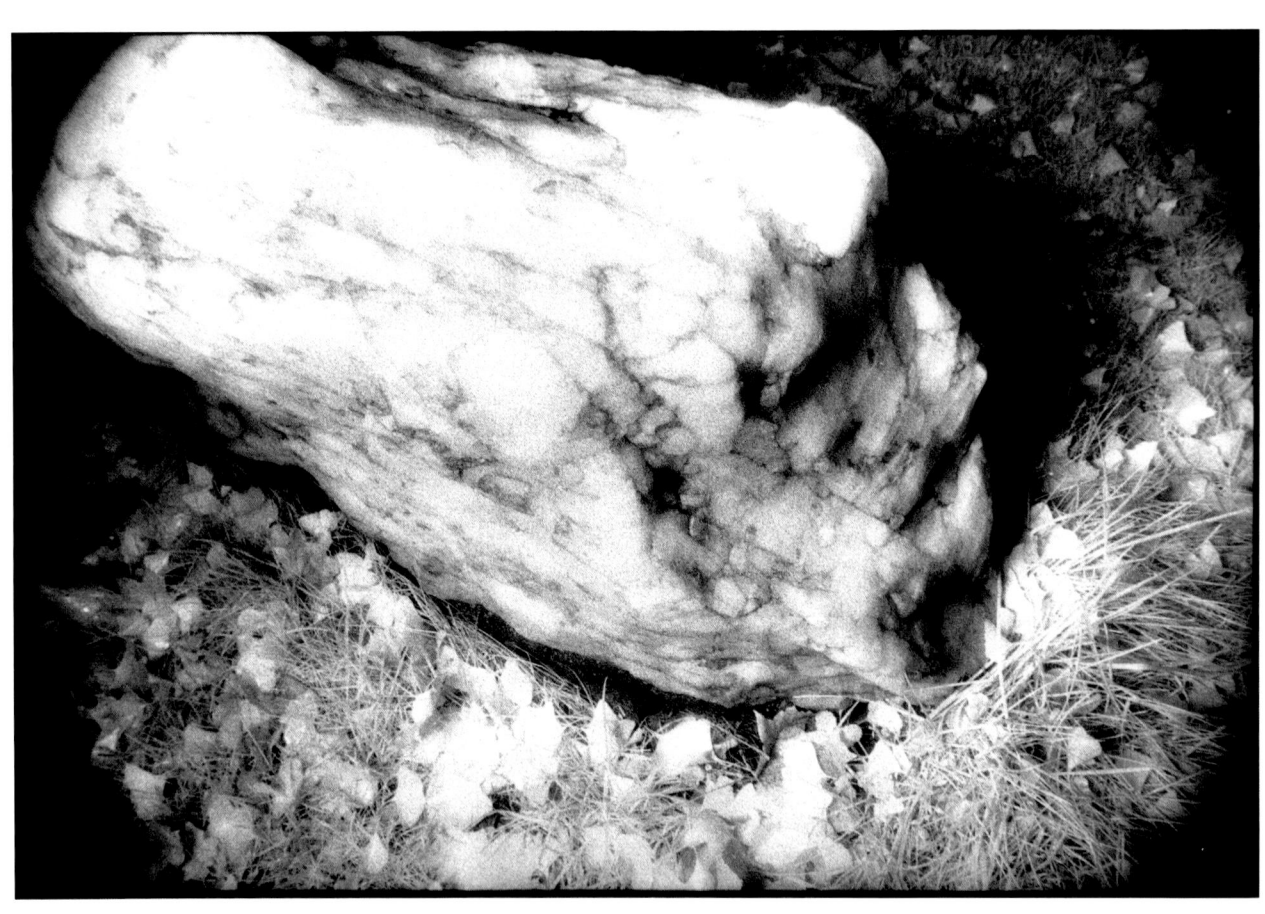

DAVID SACHTER
Rochester, New York
Rock

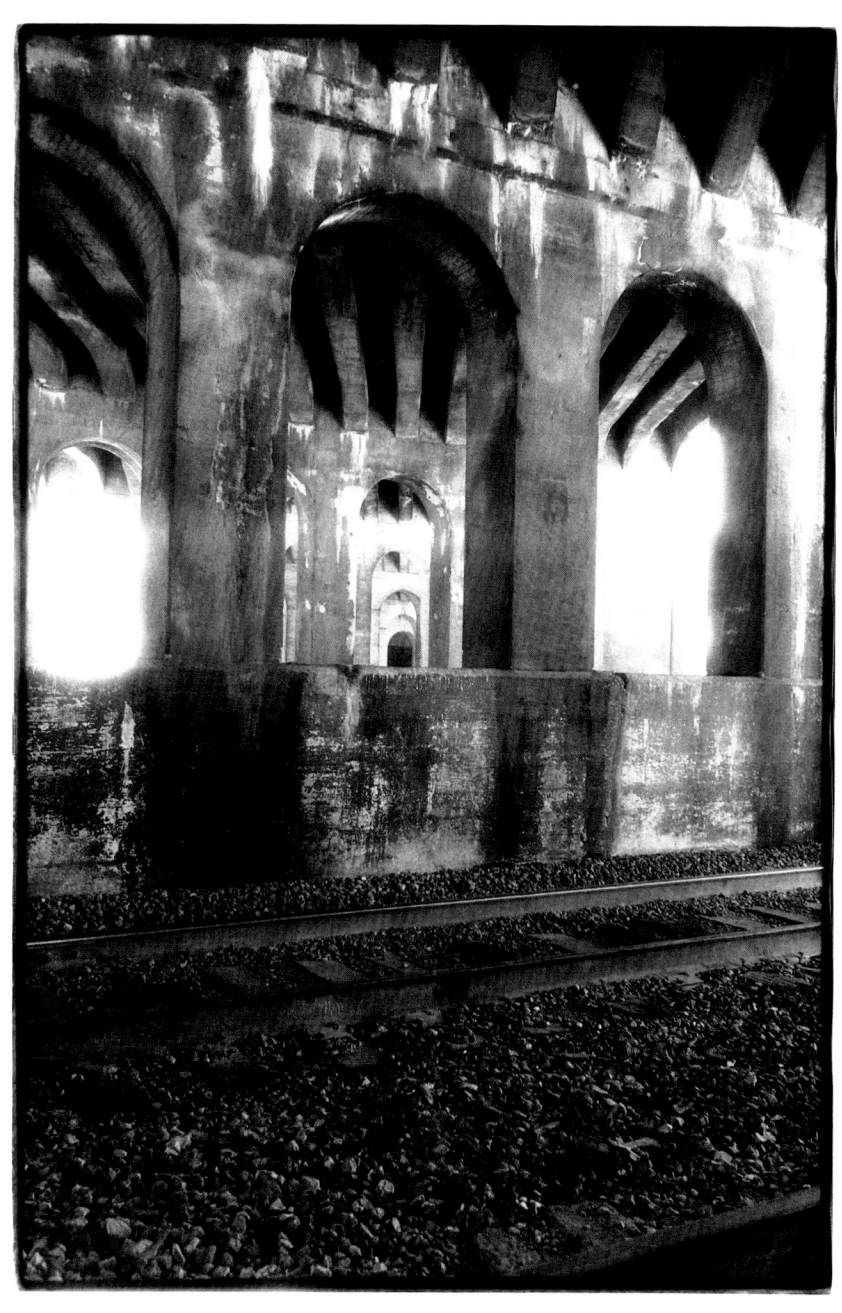

KRISTIN MUELLER READ
Ann Arbor, Michigan
Untitled, Ann Arbor, Michigan

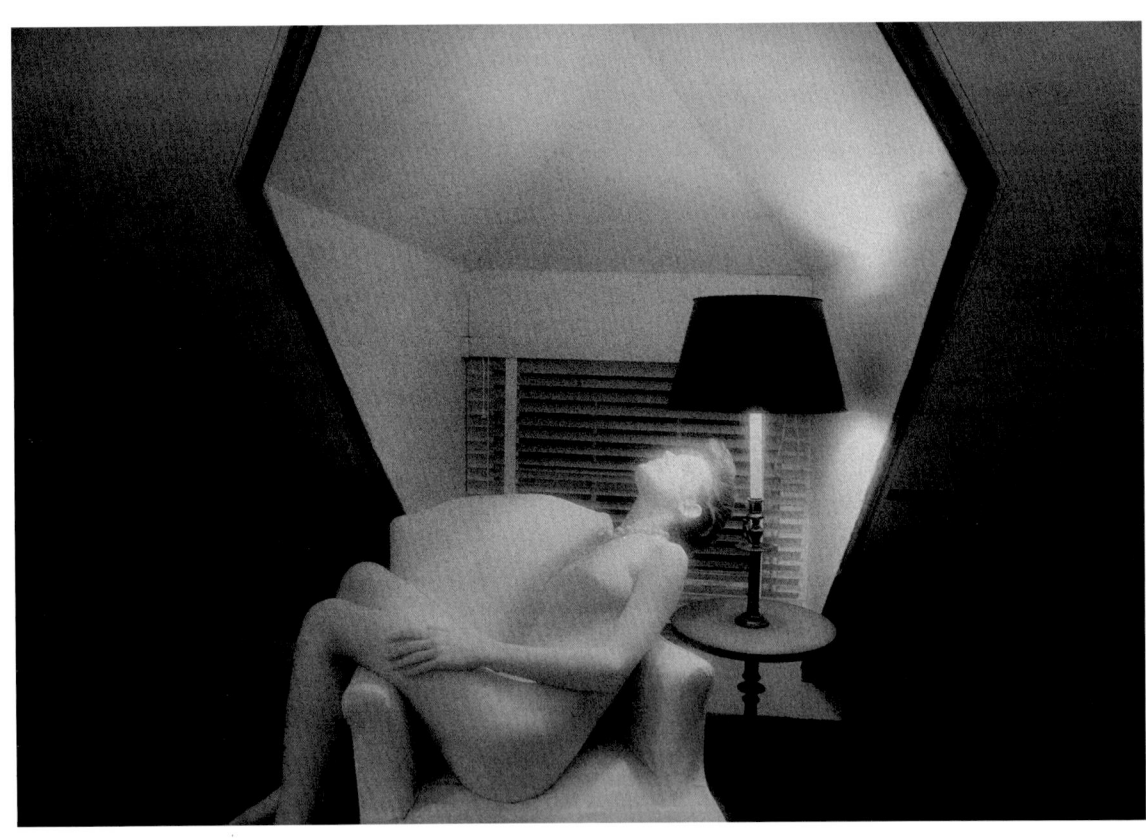

DAVID O'CONNOR
New York, New York
Barbara's Hotel Room

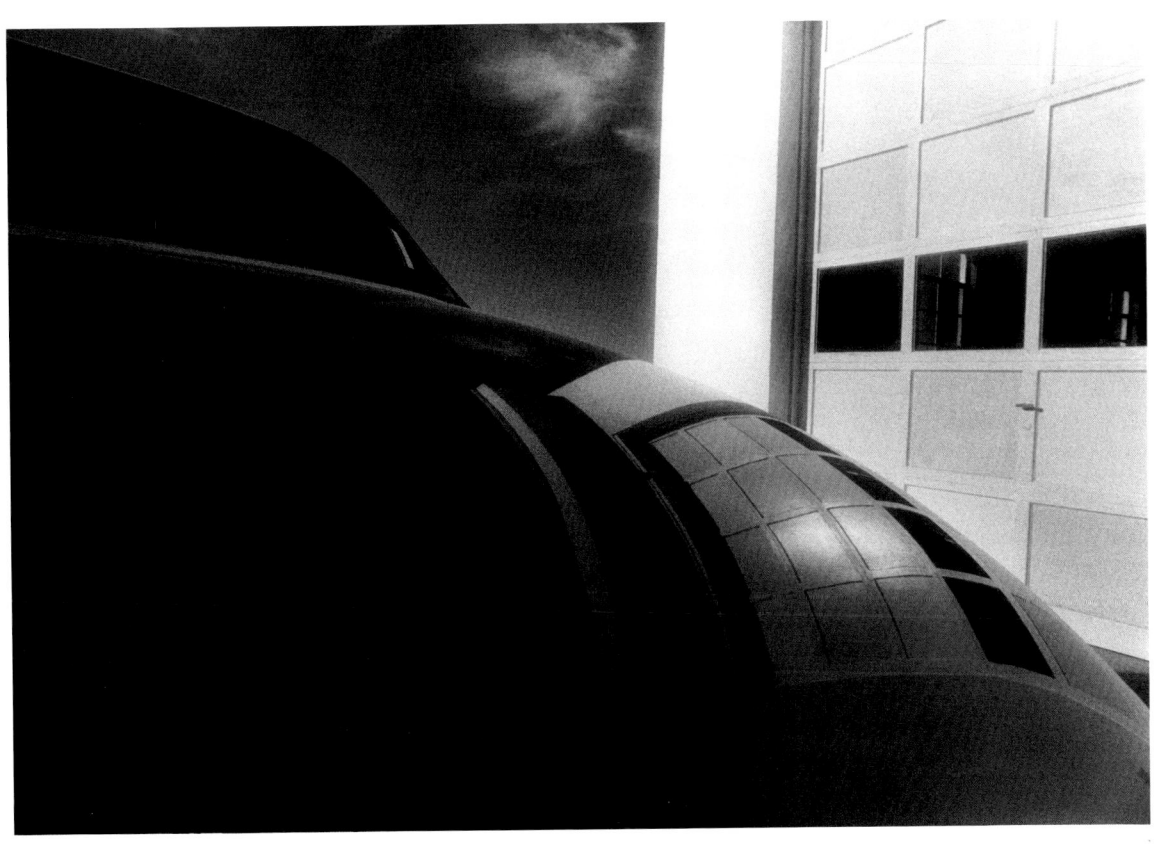

DAVID O'CONNOR
New York, New York
Automobile, 1979

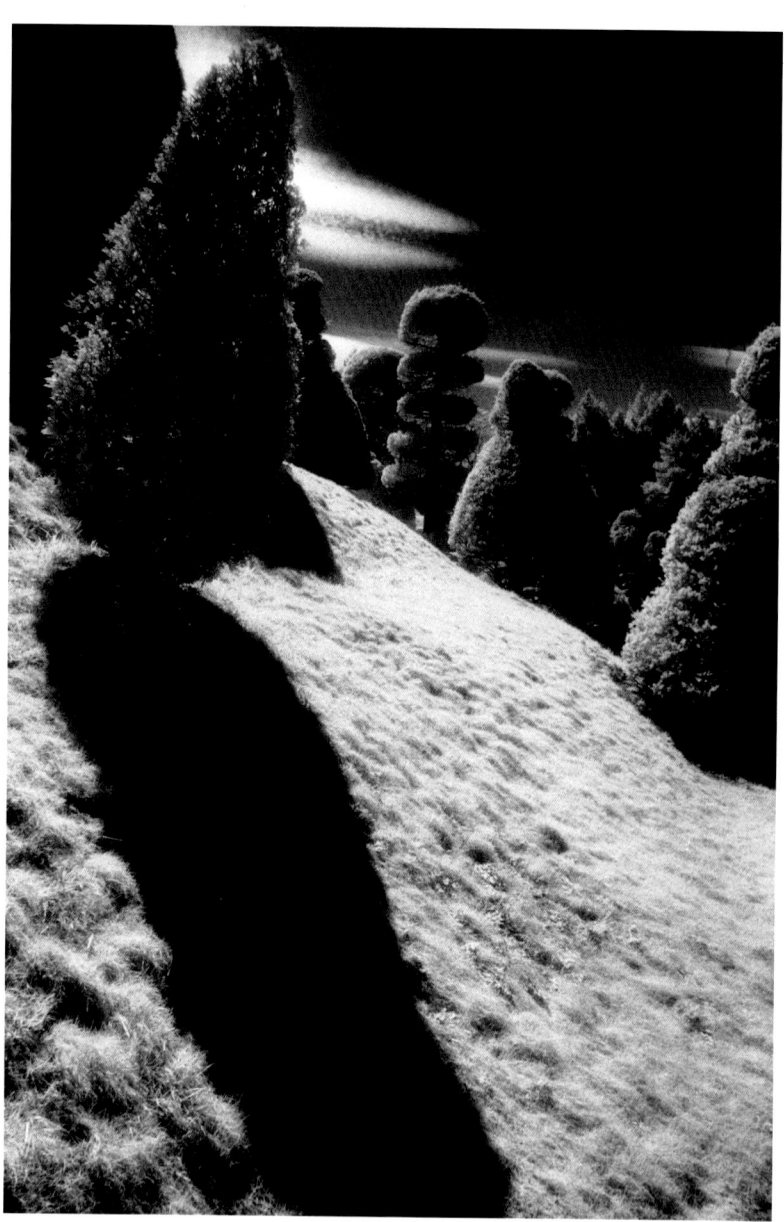

DAVID O'CONNOR
New York, New York
Topiary, 1977

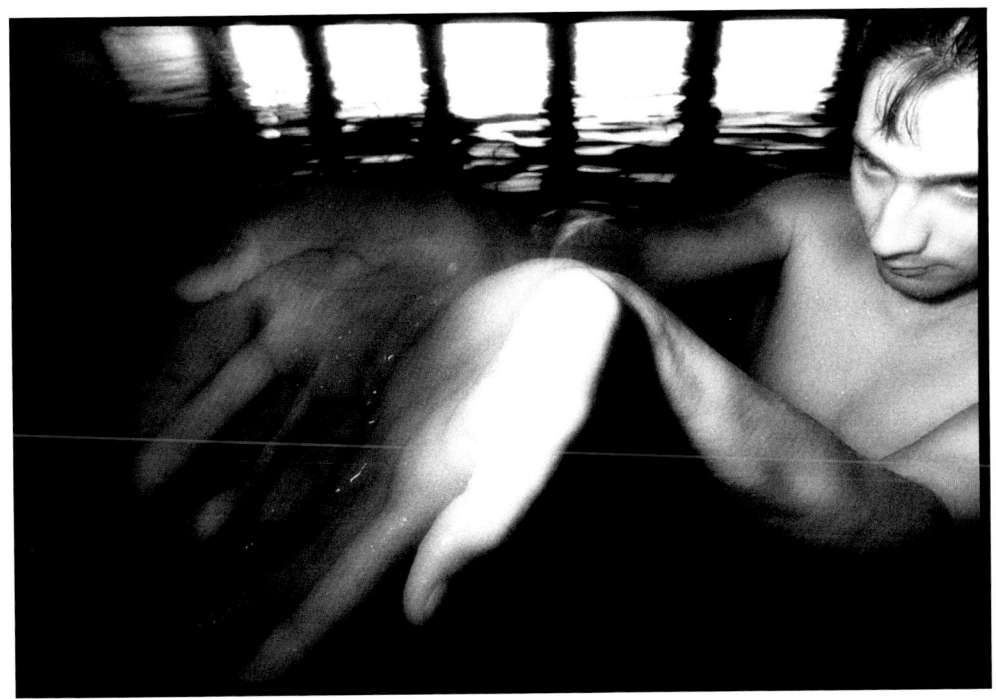

PHILIP SCHERTZ
Bryan, Texas
Slipping Into Dark, 1982 (© Philip Schertz, 1982)

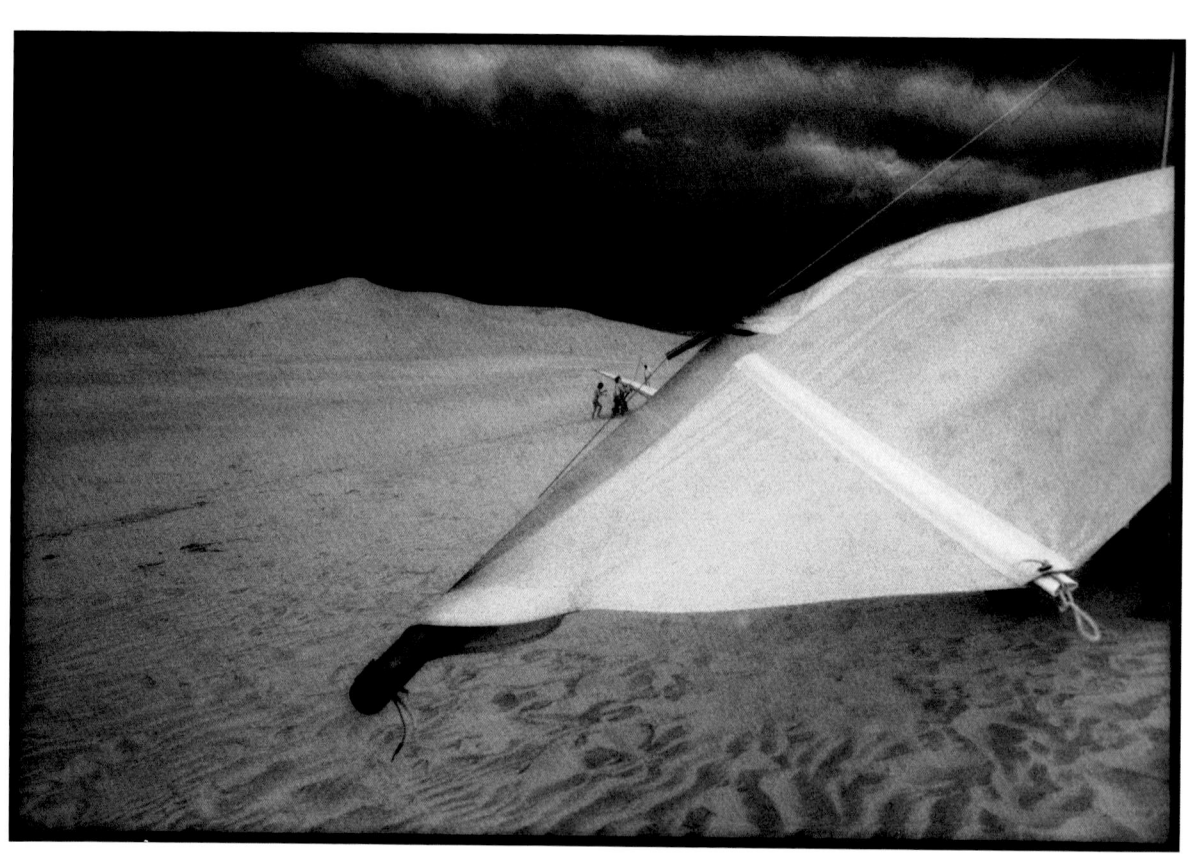

CRAIGIE HUSTON HEMENWAY
Oswego, New York
Jacob's Roost

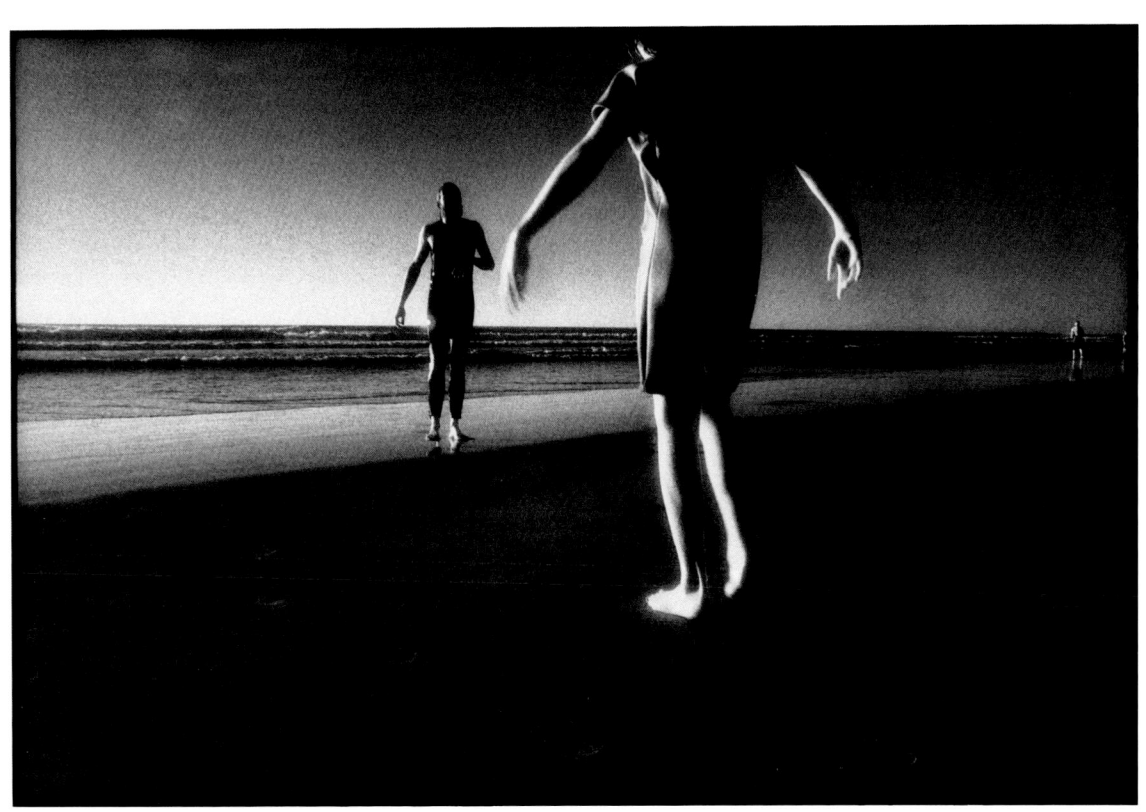

CRAIGIE HUSTON HEMENWAY
Oswego, New York
Untitled

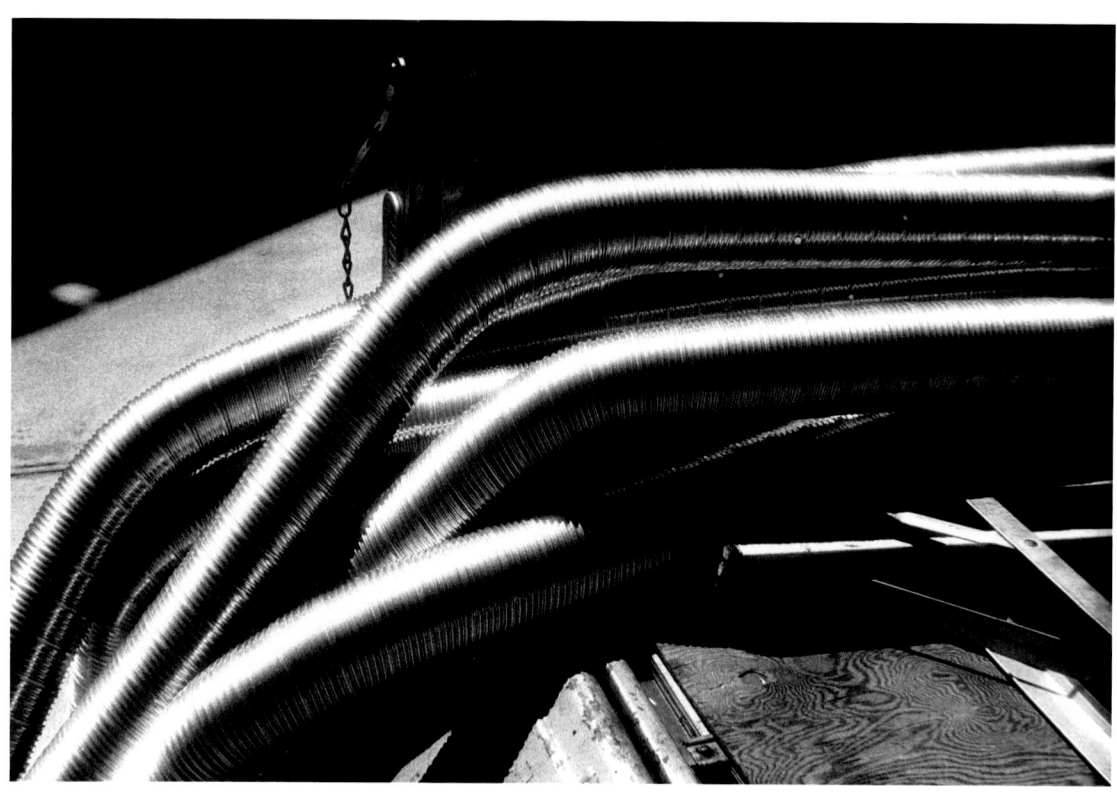

PAUL E. DAVIS
Simsbury, Connecticut

Rodondo Beach, California #2

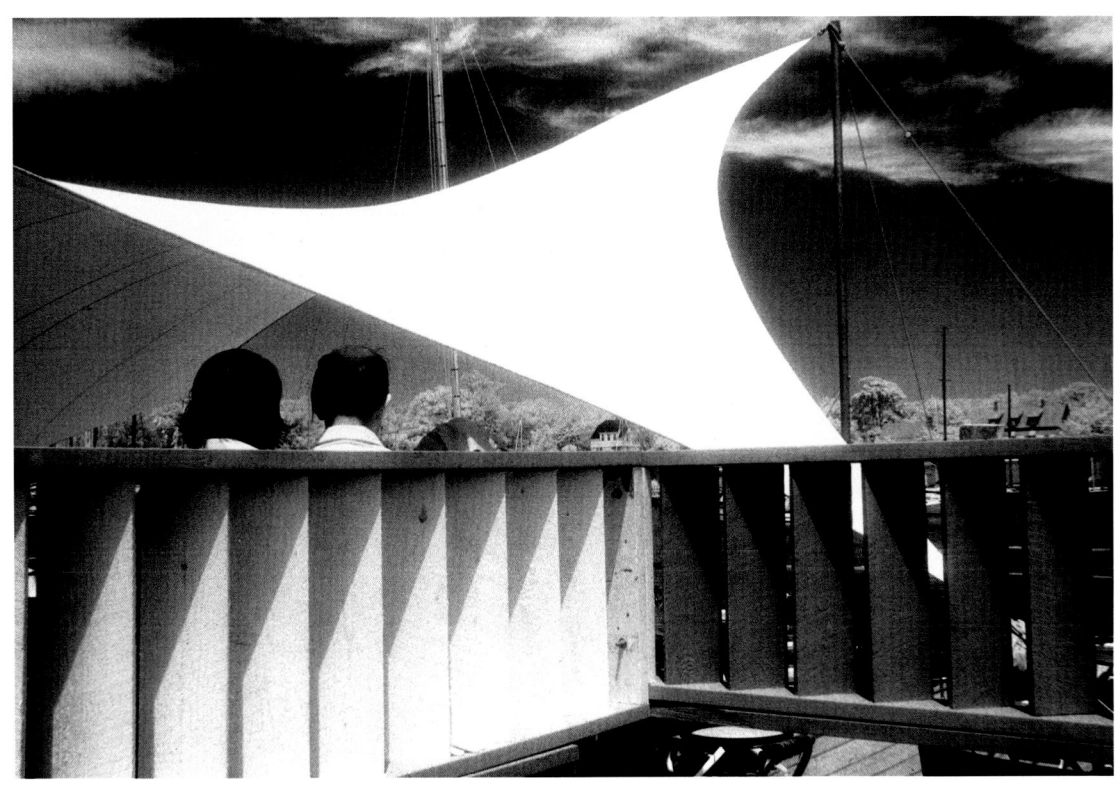

PAUL E. DAVIS
Simsbury, Connecticut
The Waterfront

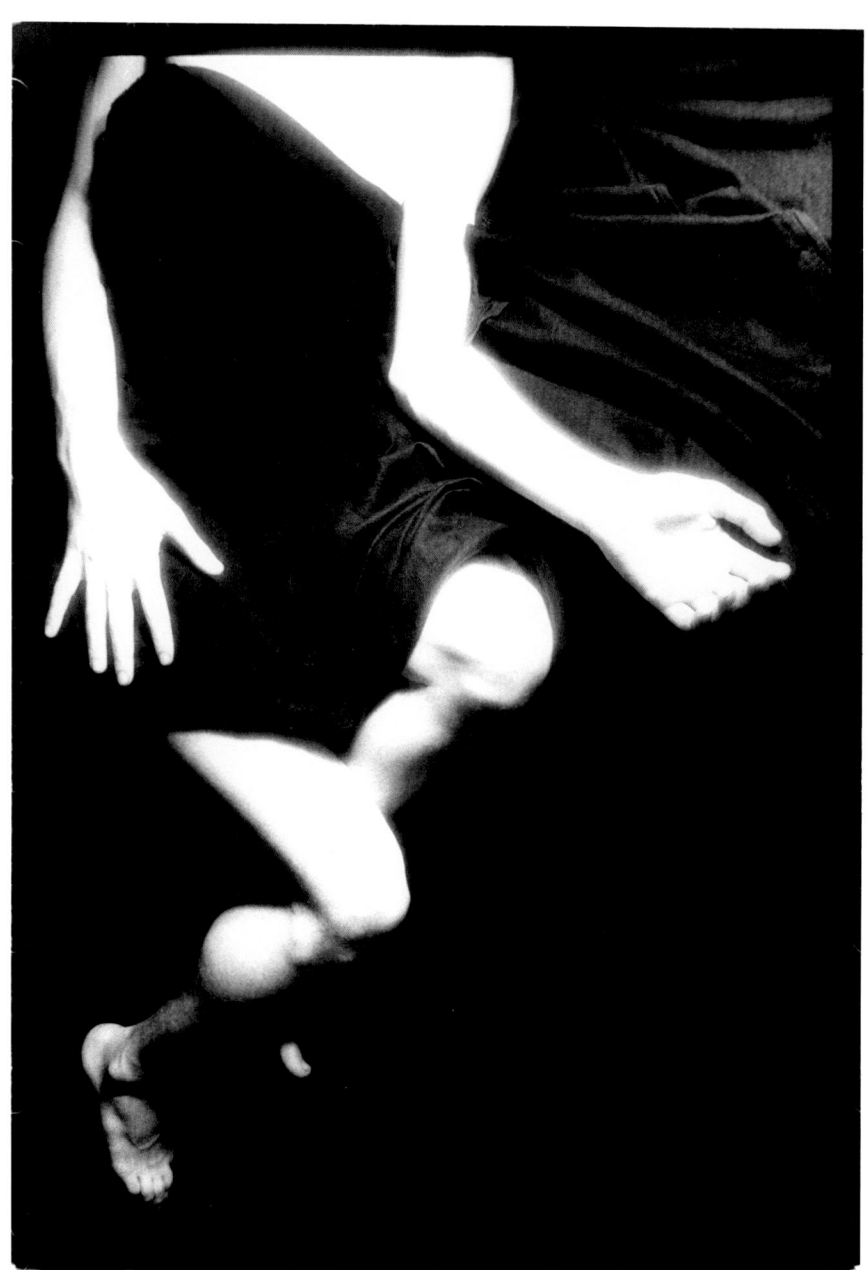

MIKE VAZQUEZ
Claremont, California
David

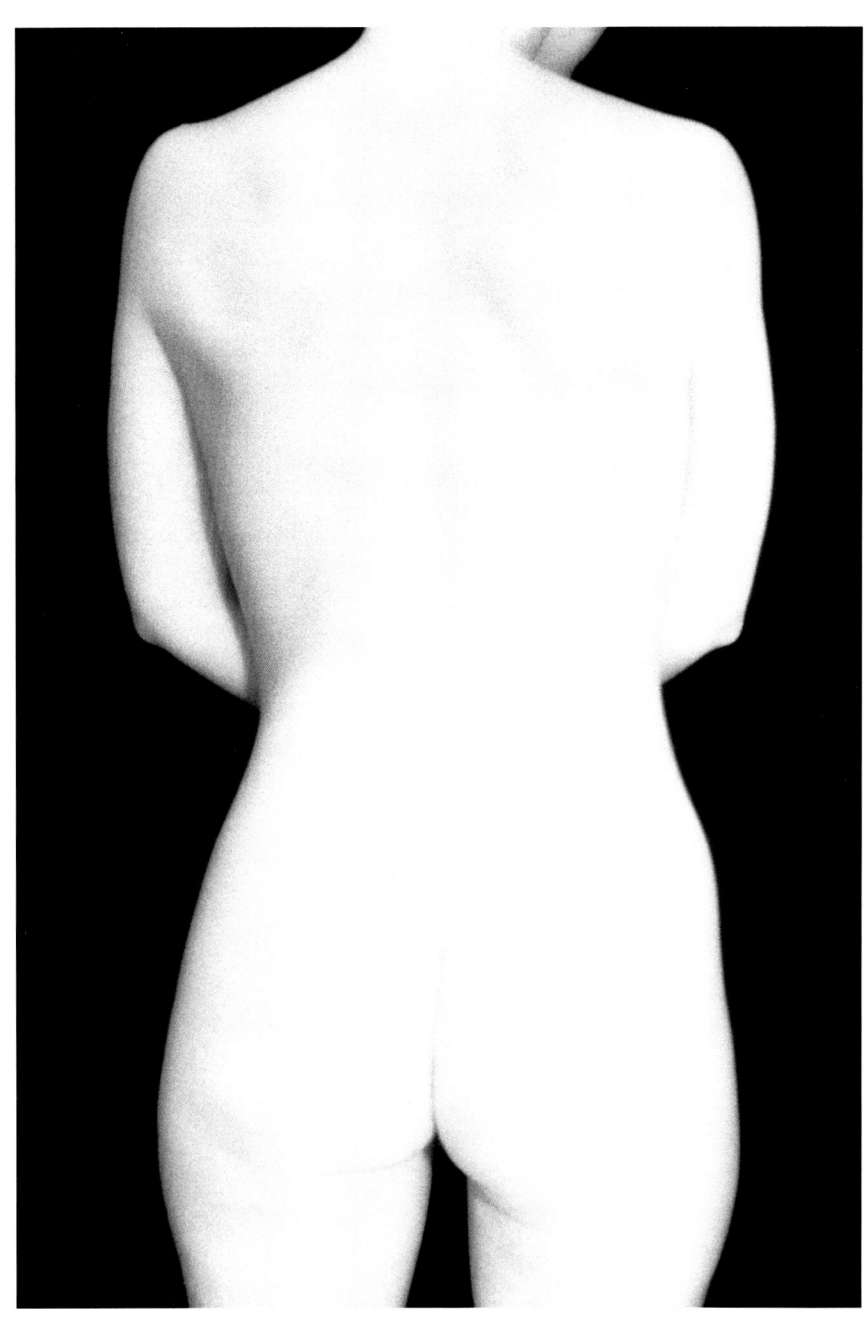

GENEVIEVE HARM
Pittsburgh, Pennsylvania
Untitled

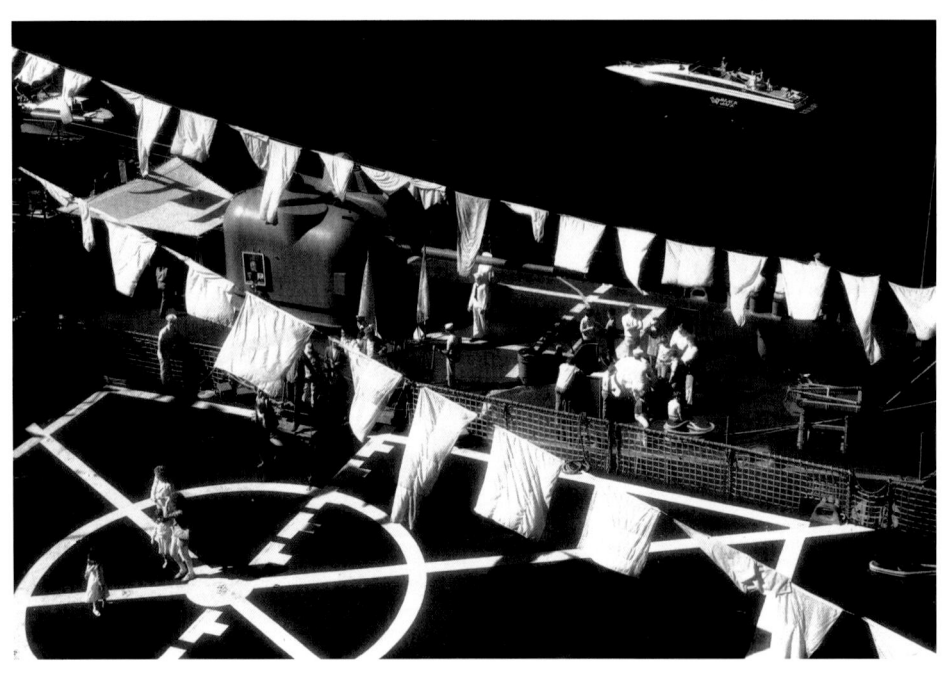

CHERYL BUSBEE
New York, New York
Black Jack, 1982

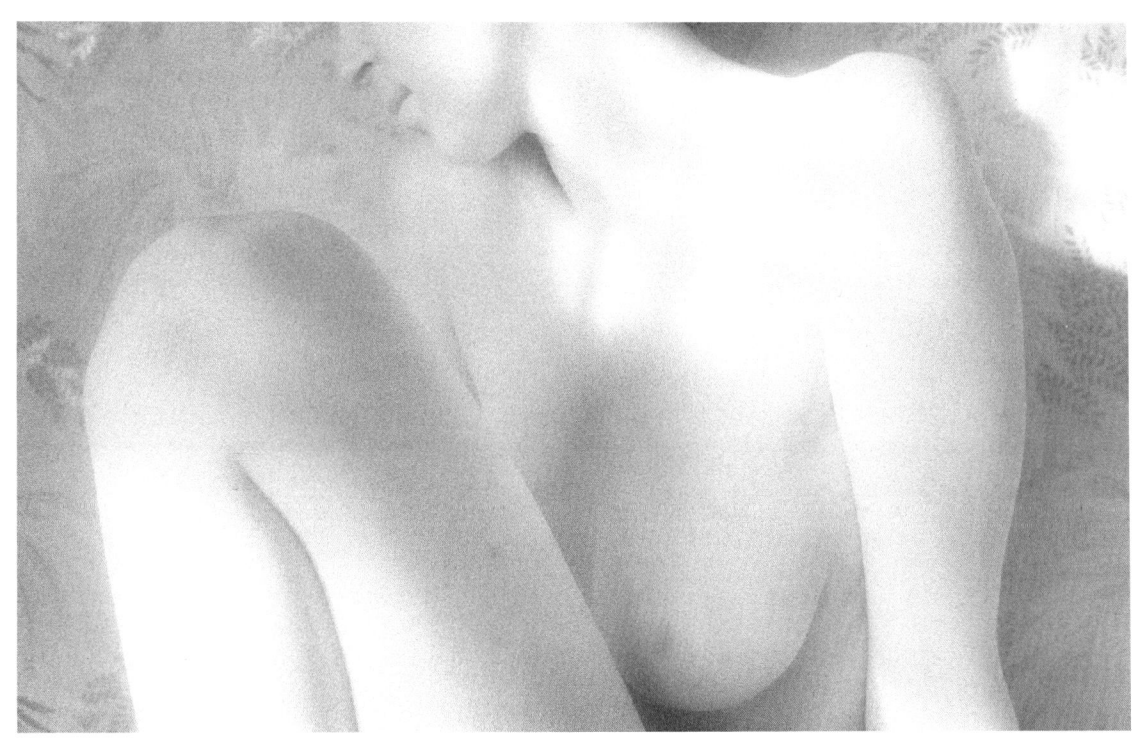

CRAIG LAW
Logan, Utah
Untitled

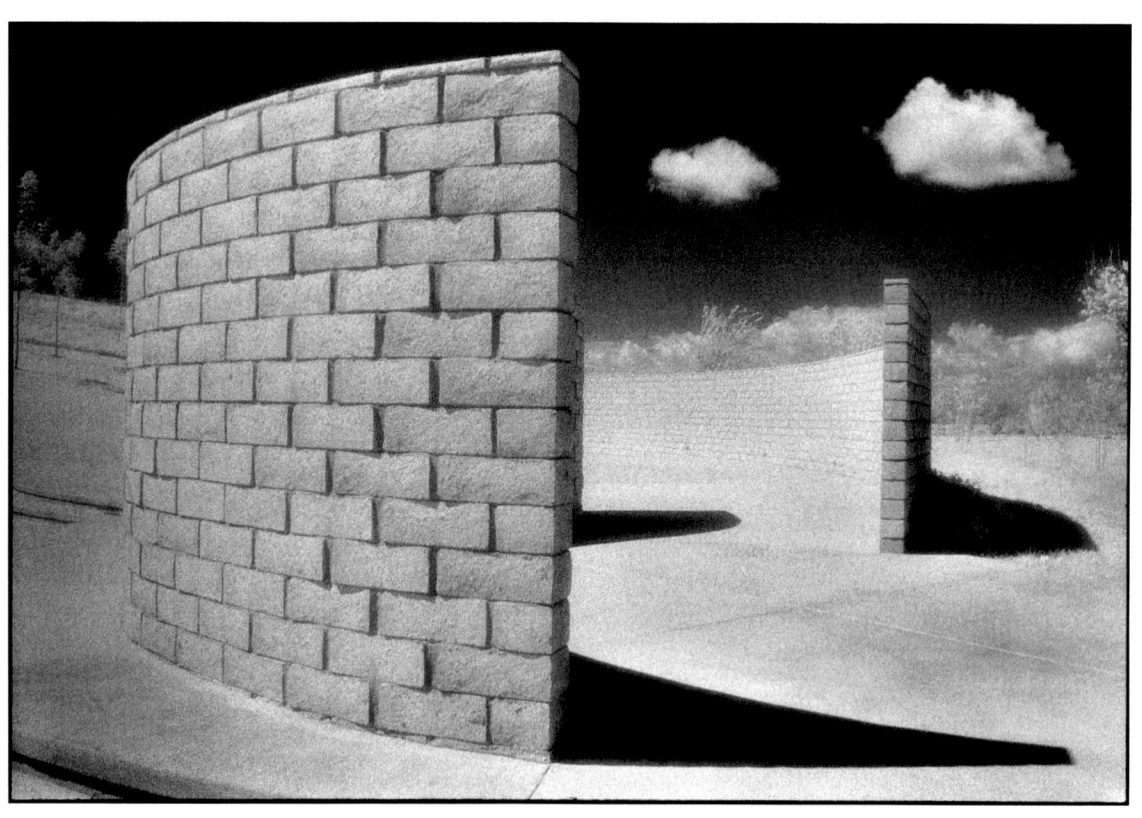

RICK FERNCASE
Costa Mesa, California
Suburban Landscape

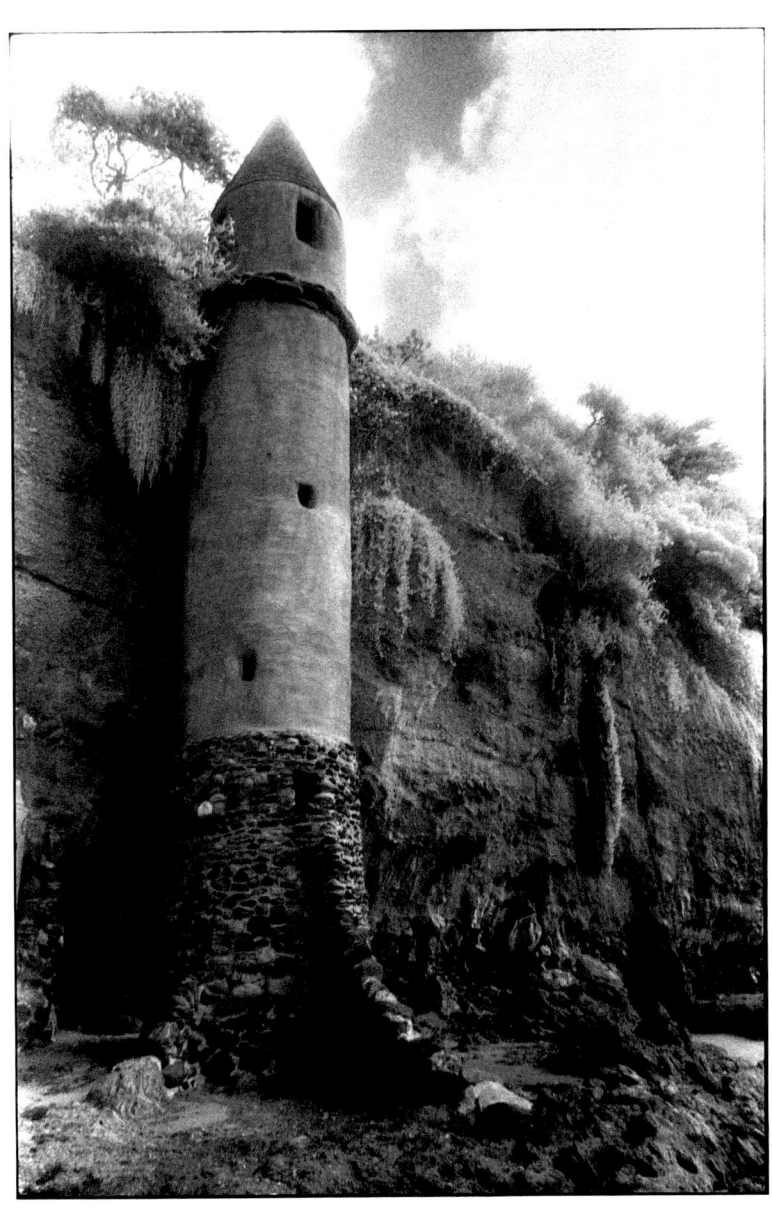

RICK FERNCASE
Costa Mesa, California
Any Of Certain Lofty Towers

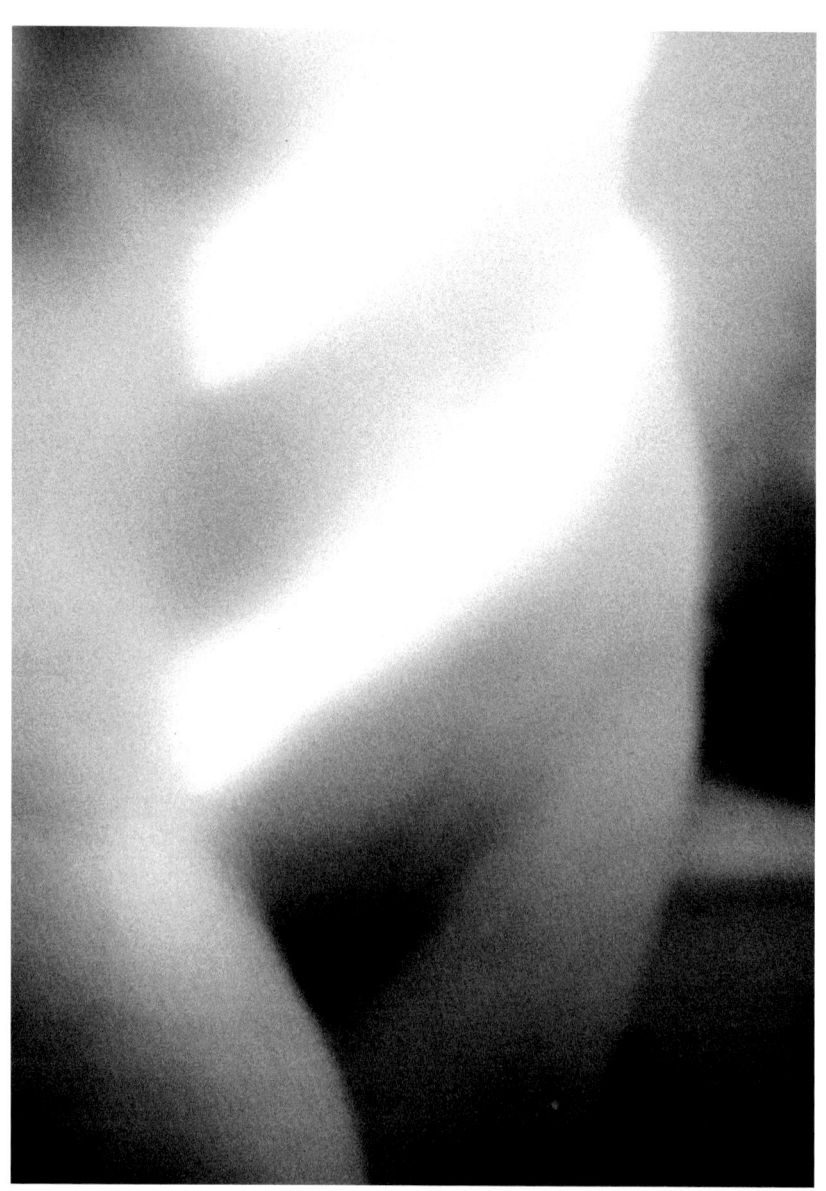

RICHARD ALBRIGHT
Pocatello, Idaho
Jodeane #1, 1980

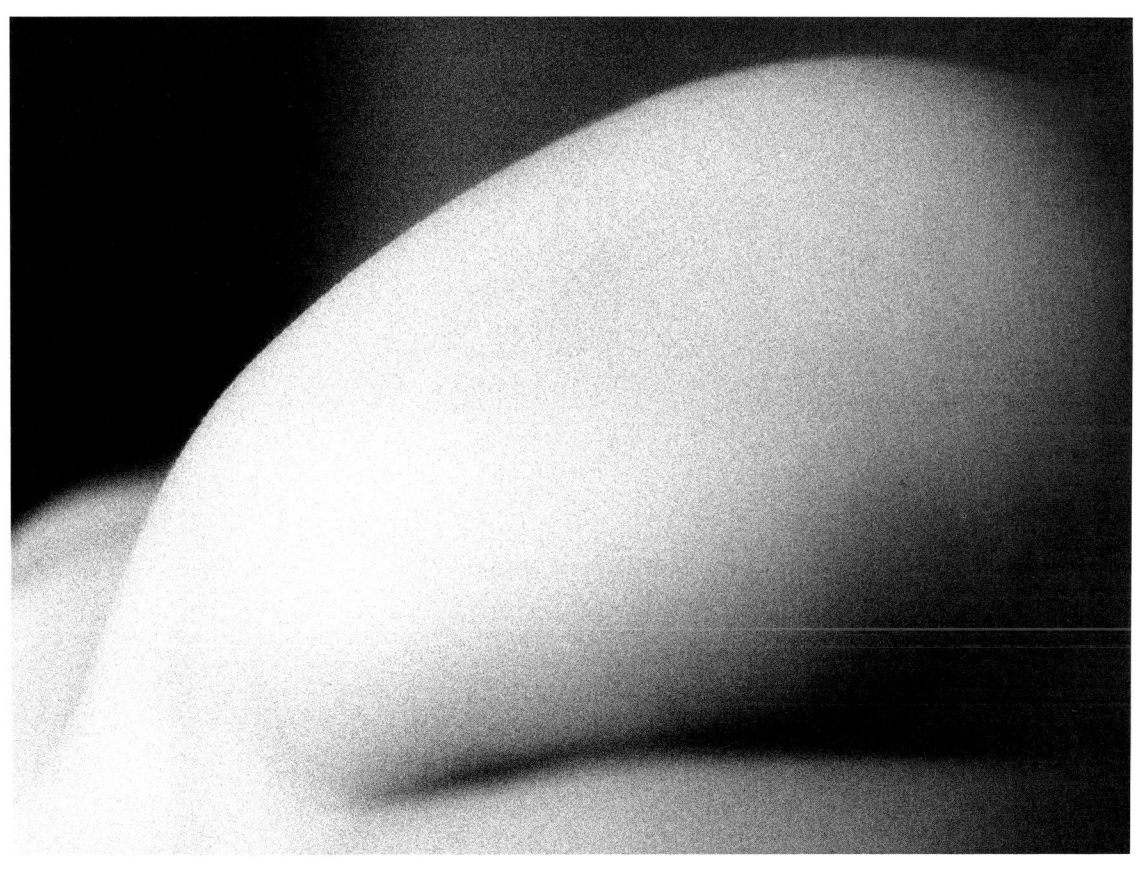

RICHARD ALBRIGHT
Pocatello, Idaho
Silent Sleep, 1980

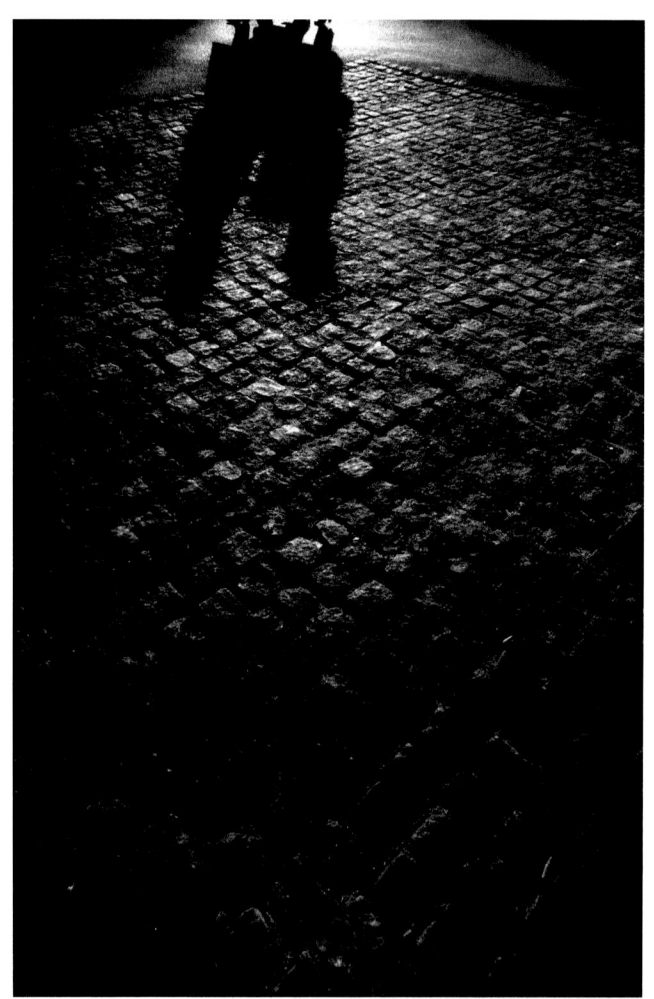

ROY J. CIRIGLIANA
Dallas, Texas
Untitled

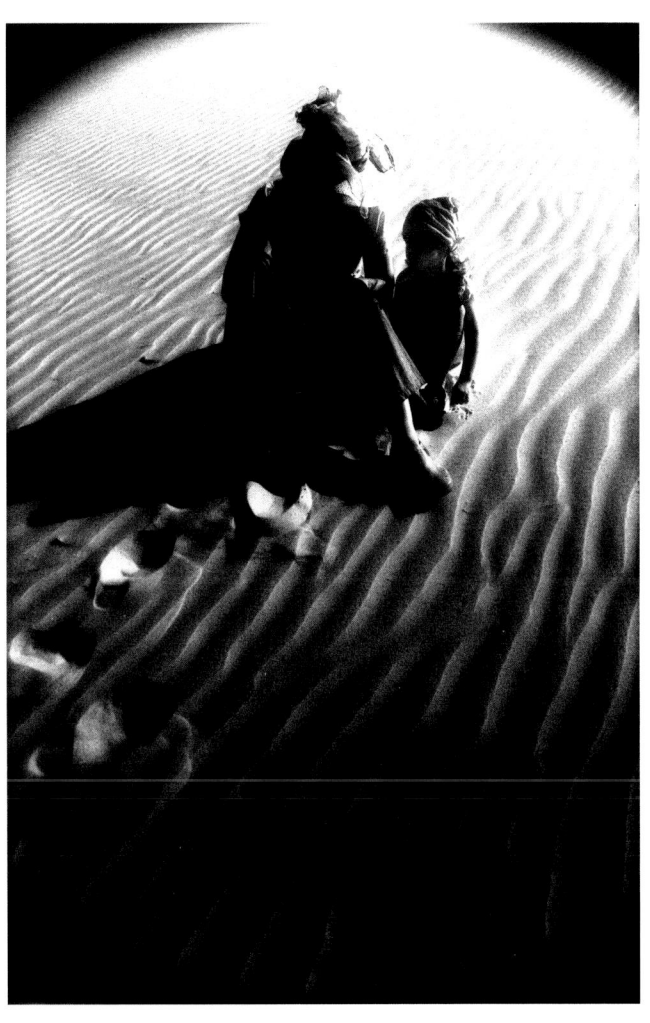

ROY J. CIRIGLIANA
Dallas, Texas
Untitled

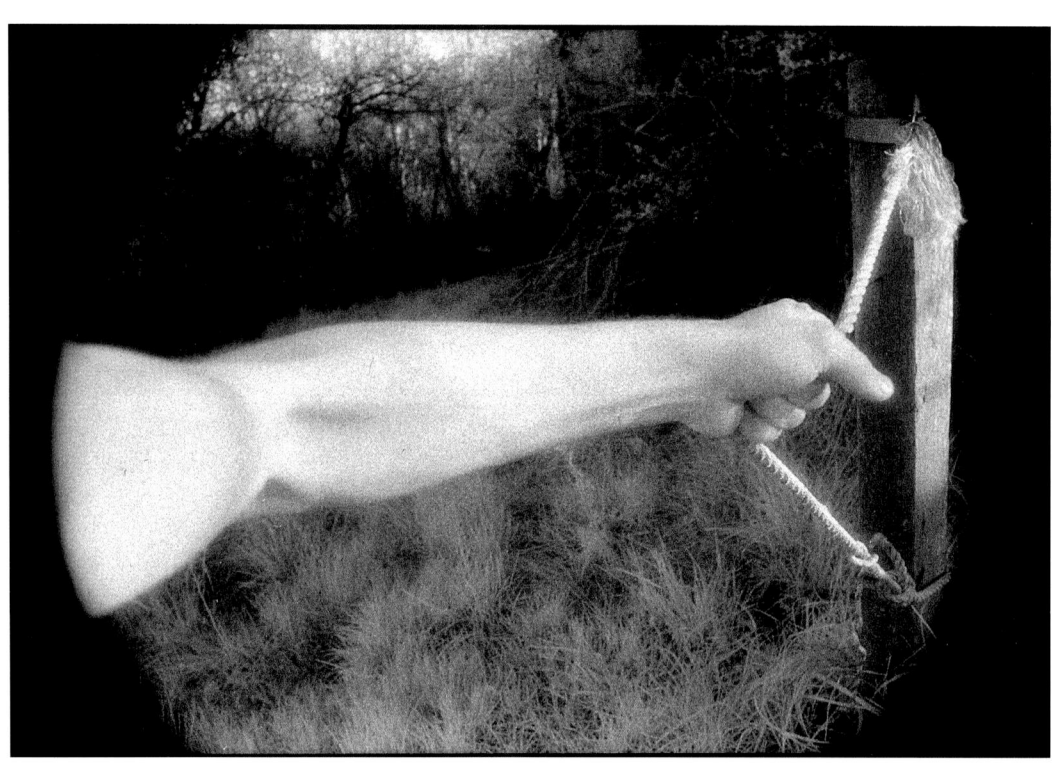

JEFFERY NEWBURY
San Francisco, California
Untitled

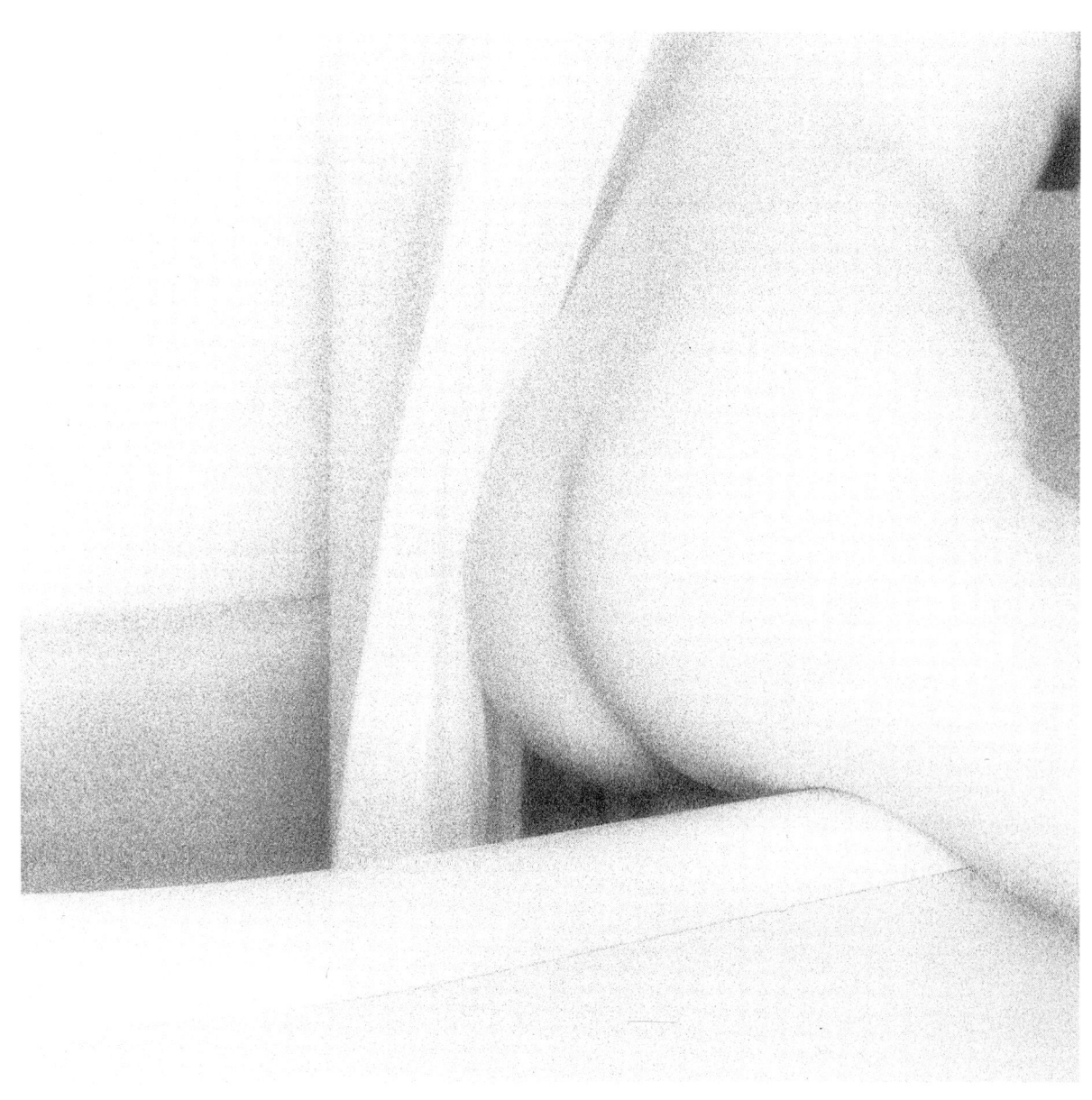

NANCY P. ROBERTS
Milton, Massachusetts
Self Portrait

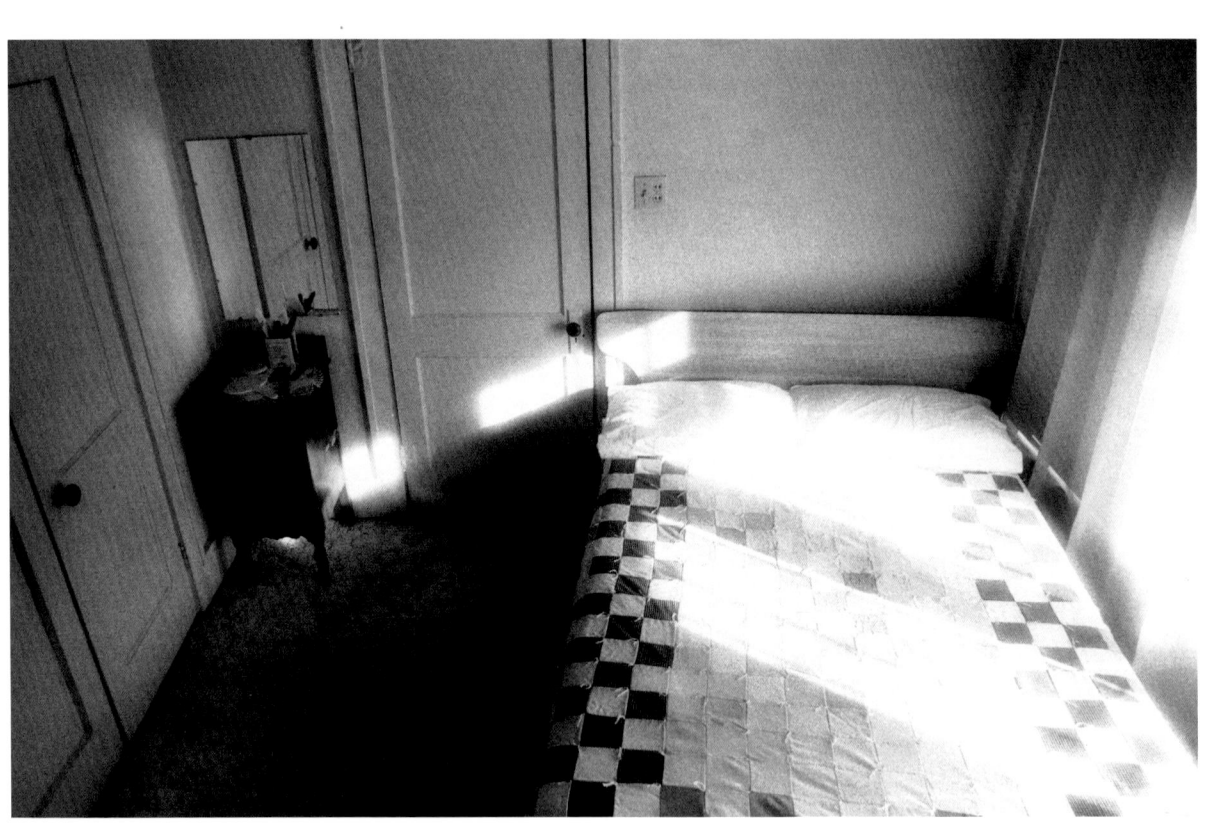

MARK L. KETTER
Elkgrove, Illinois

No. 19, Morning Weeping On A Blanket Made By Hand

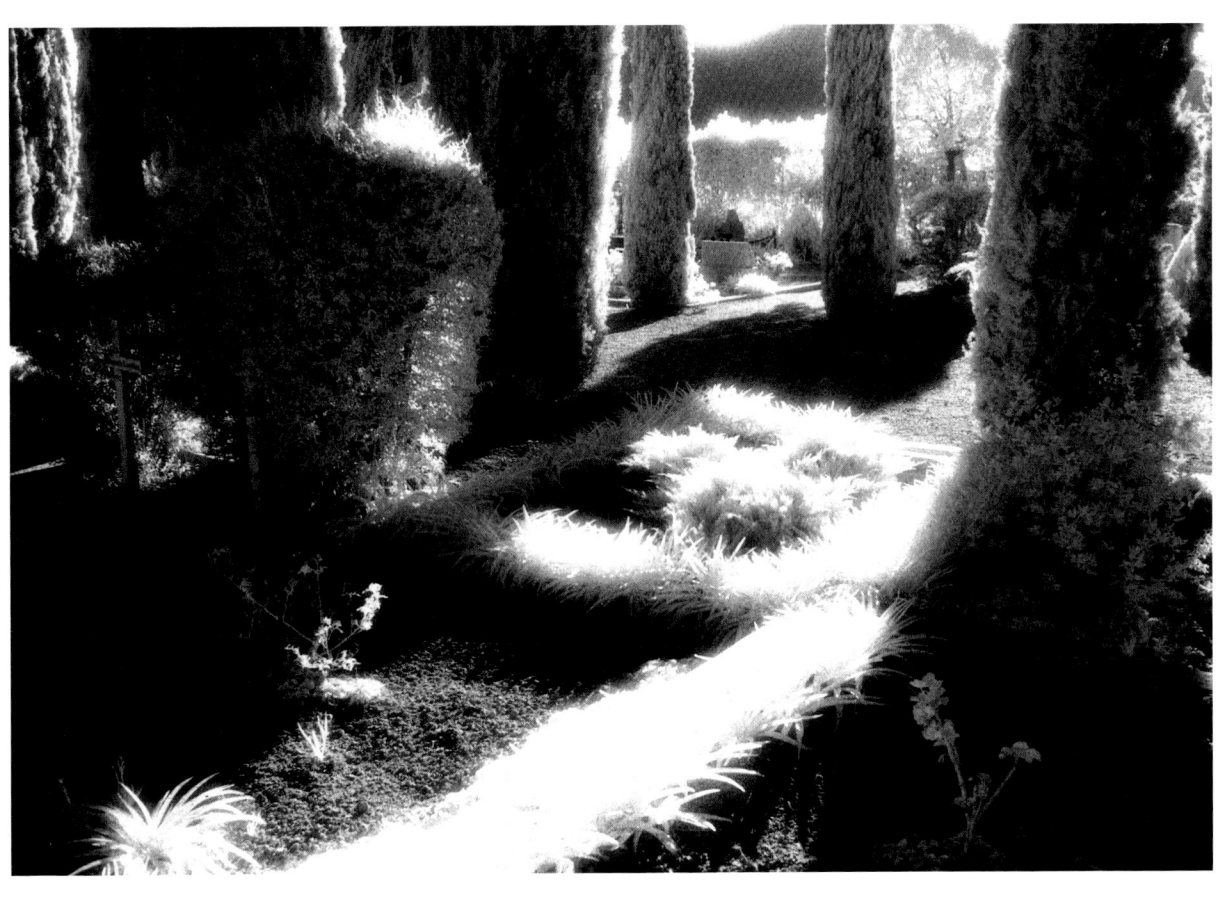

MARK L. KETTER
Elkgrove, Illinois
Number 37, Untitled

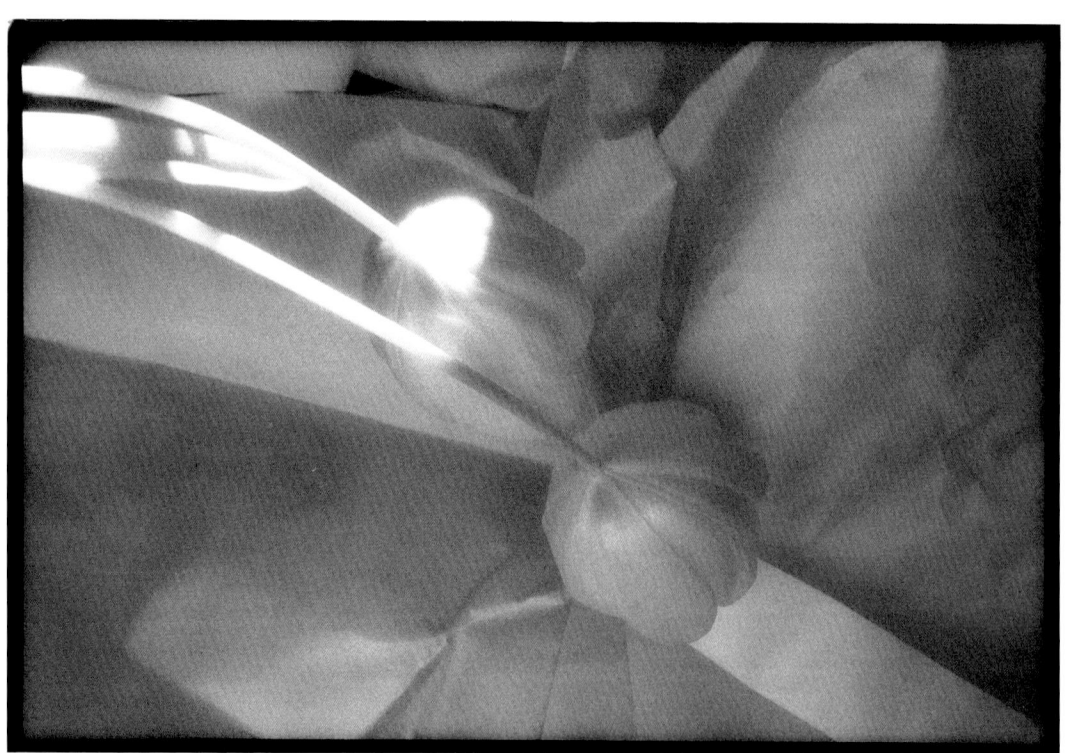

MARGUERITE RULE JOHNSTONE
New York, New York

"Two Tulips" From The "White Flower" Series
(courtesy: The Witkin Gallery, New York, New York)

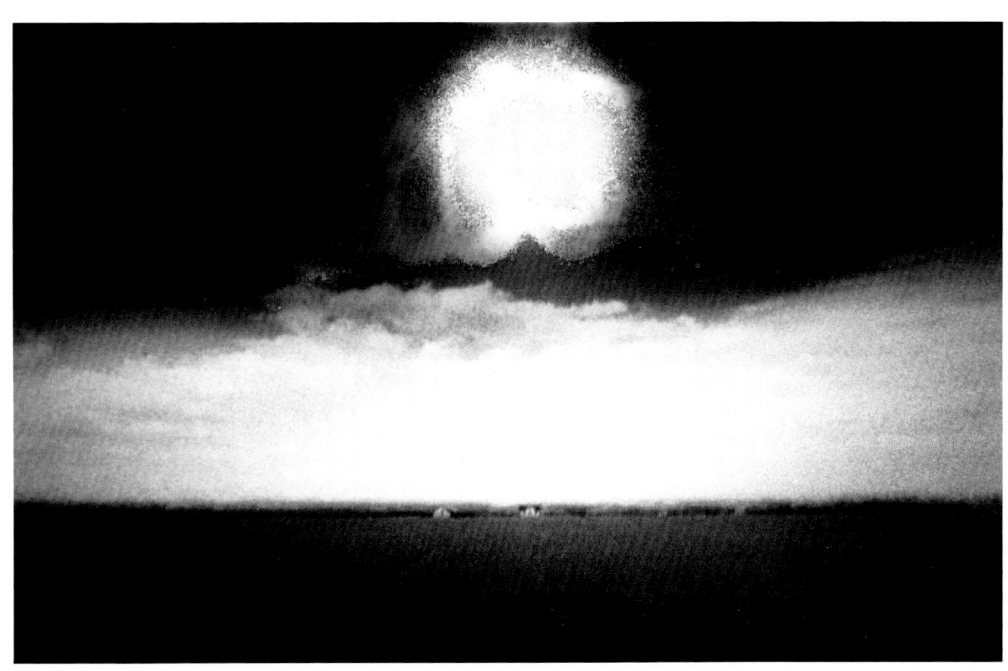

DANIEL W. ARTHUR
Waynesboro, Pennsylvania
Long Island Farmhouse

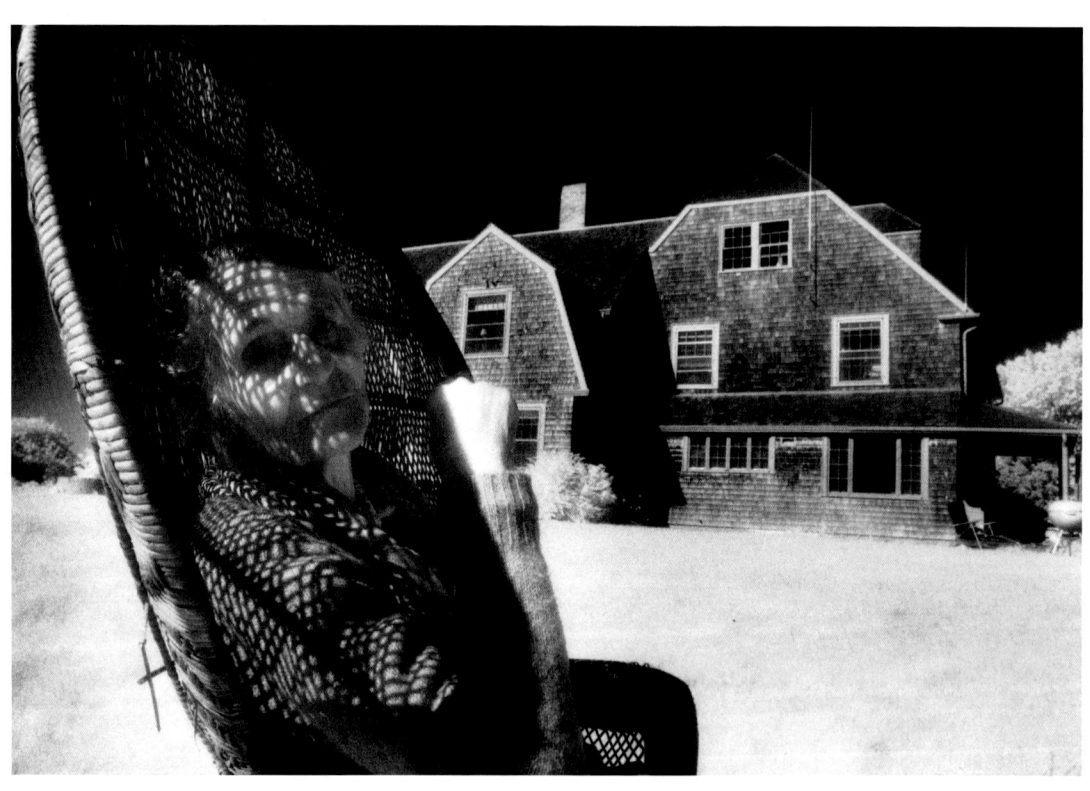

PENNY GENTIEU
New York, New York
Untitled

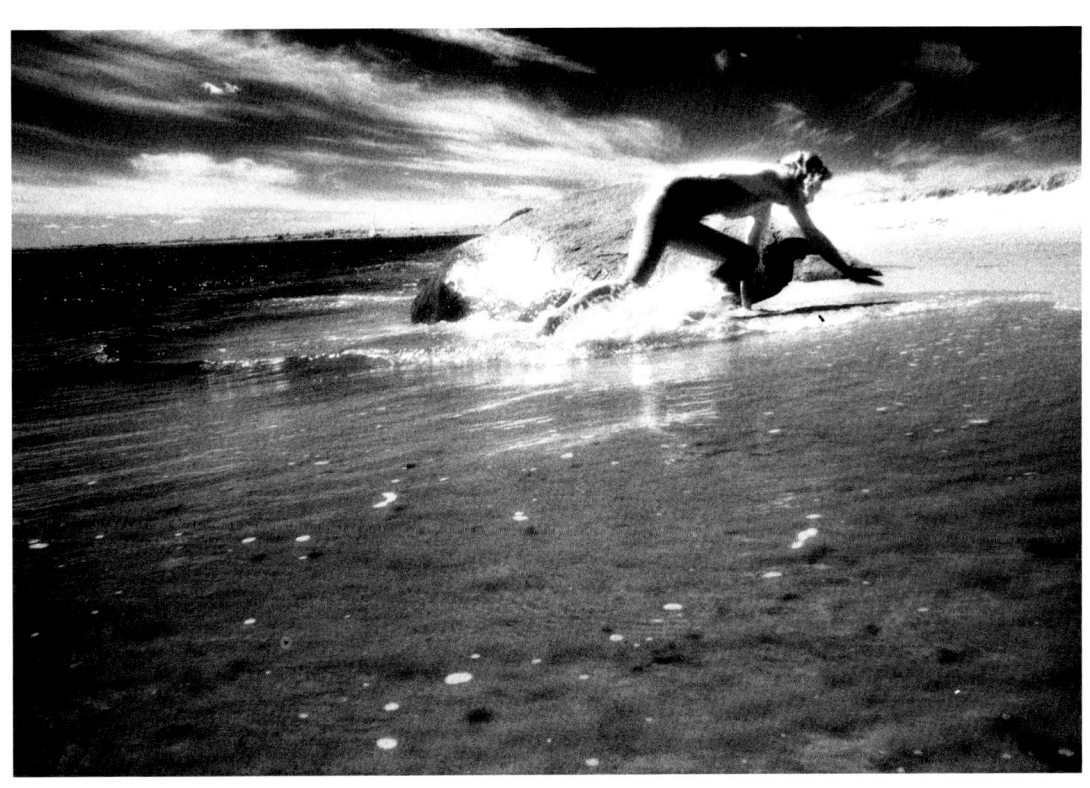

PENNY GENTIEU
New York, New York
Untitled

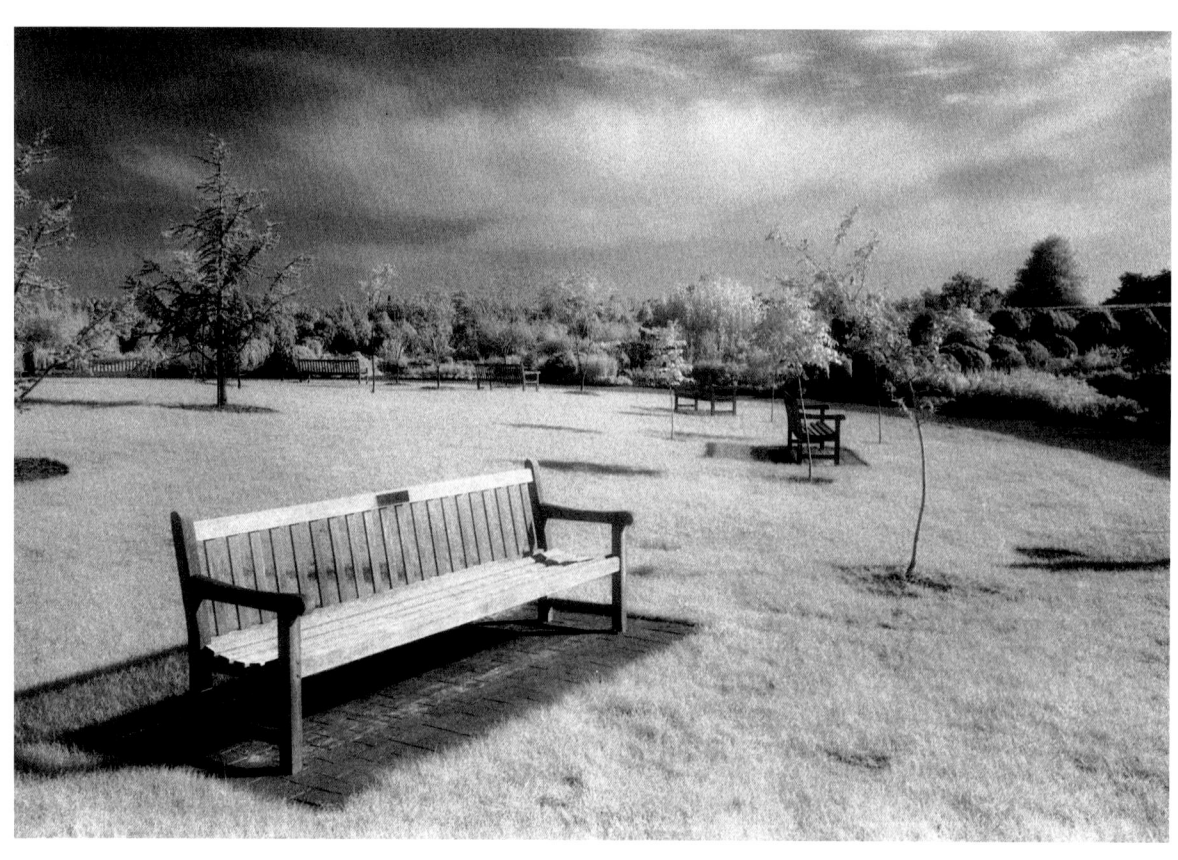

JILL ENFIELD
New York, New York
Untitled

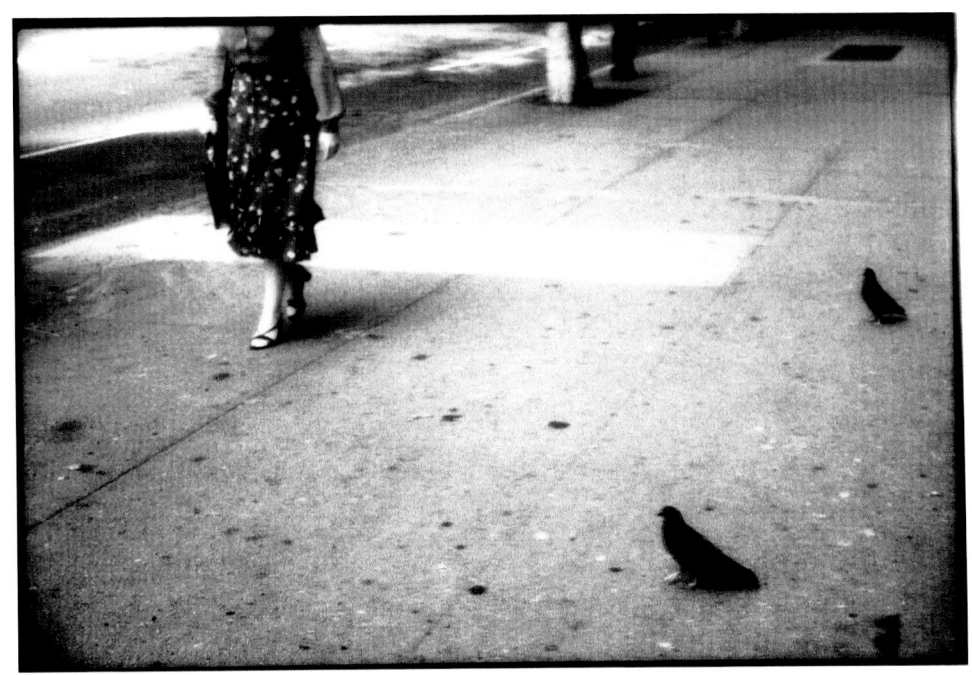

JOHN GRIFFIN
Queens Village, New York
Untitled

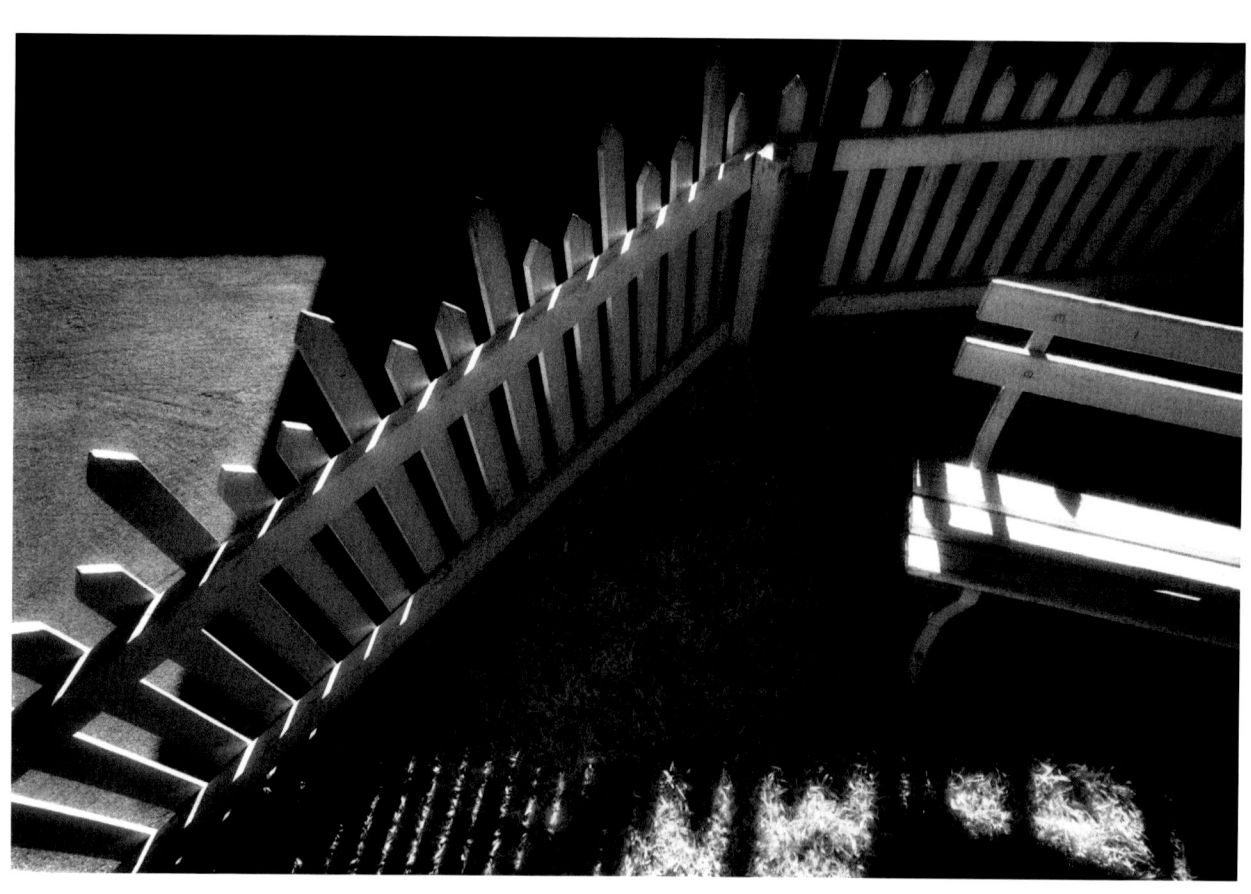

CHARLES BRAENDLE
Phoenix, Arizona
Morning Light, 1979

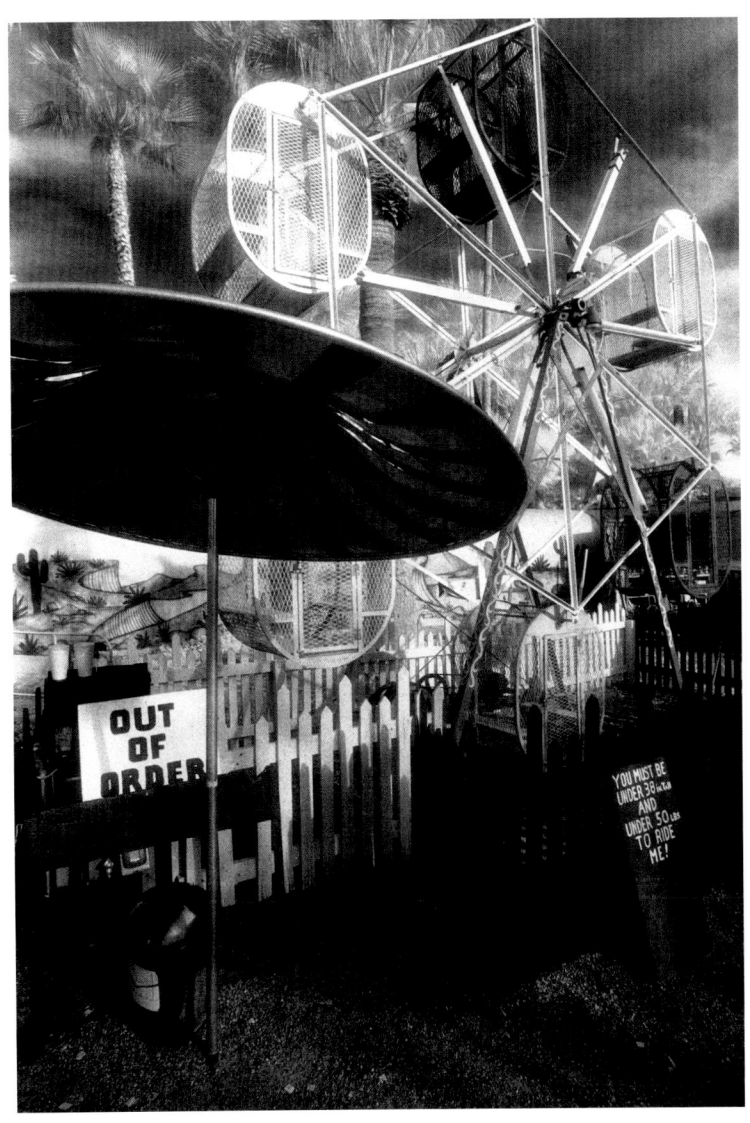

CHARLES BRAENDLE
Phoenix, Arizona
Out Of Order

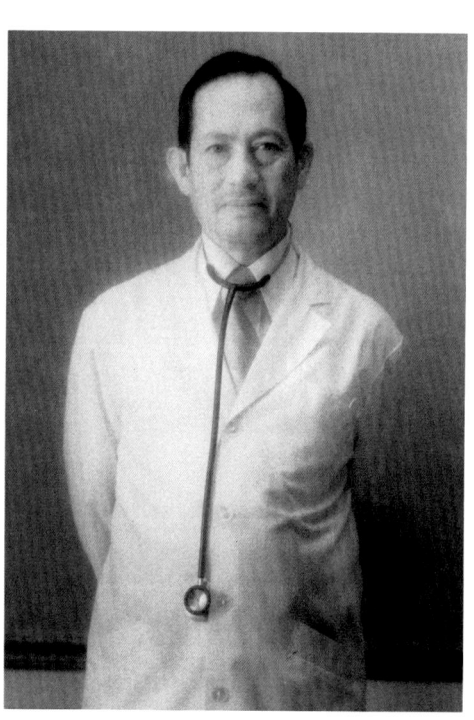
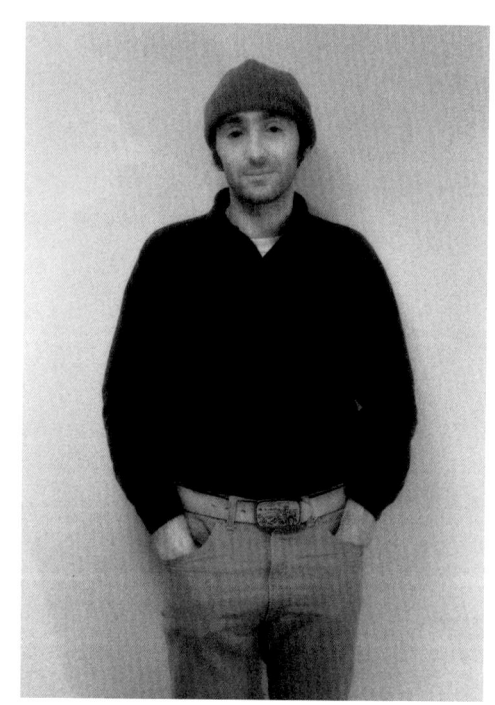

KEN MARCHIONNO
Dover, Delaware

Dr. Bañez / John

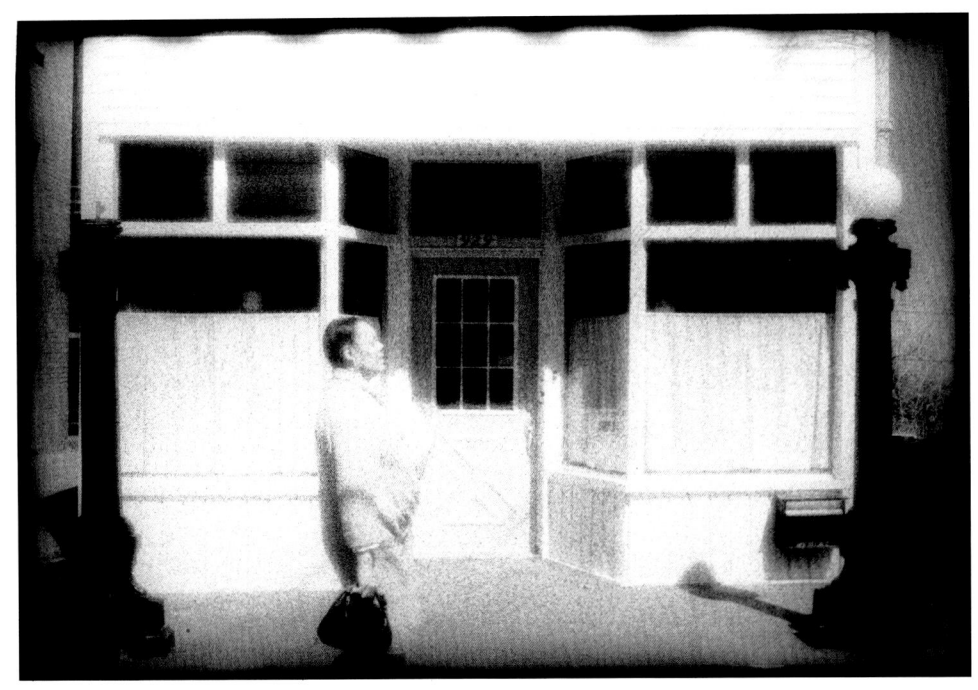

SUSAN STEIN
Brighton, Massachusetts
Untitled

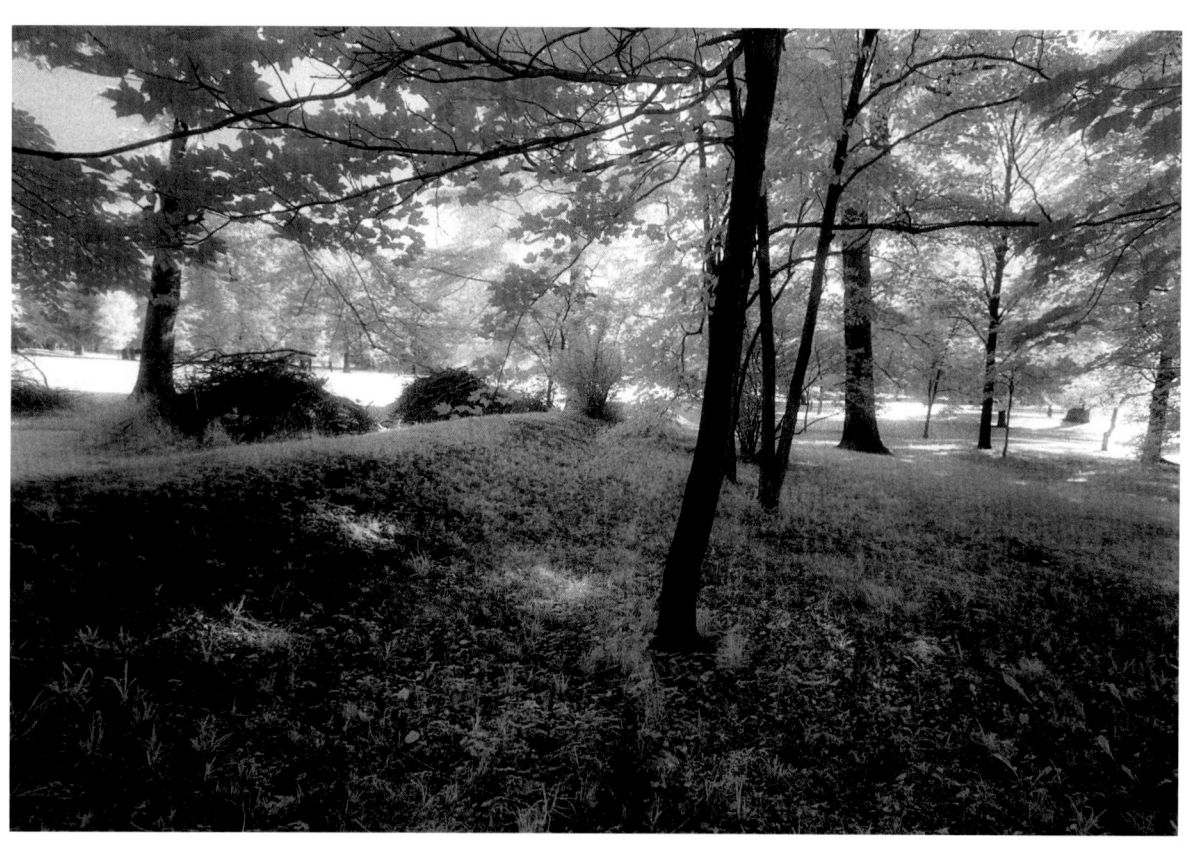

KENT BOWSER
Columbus, Ohio
Untitled

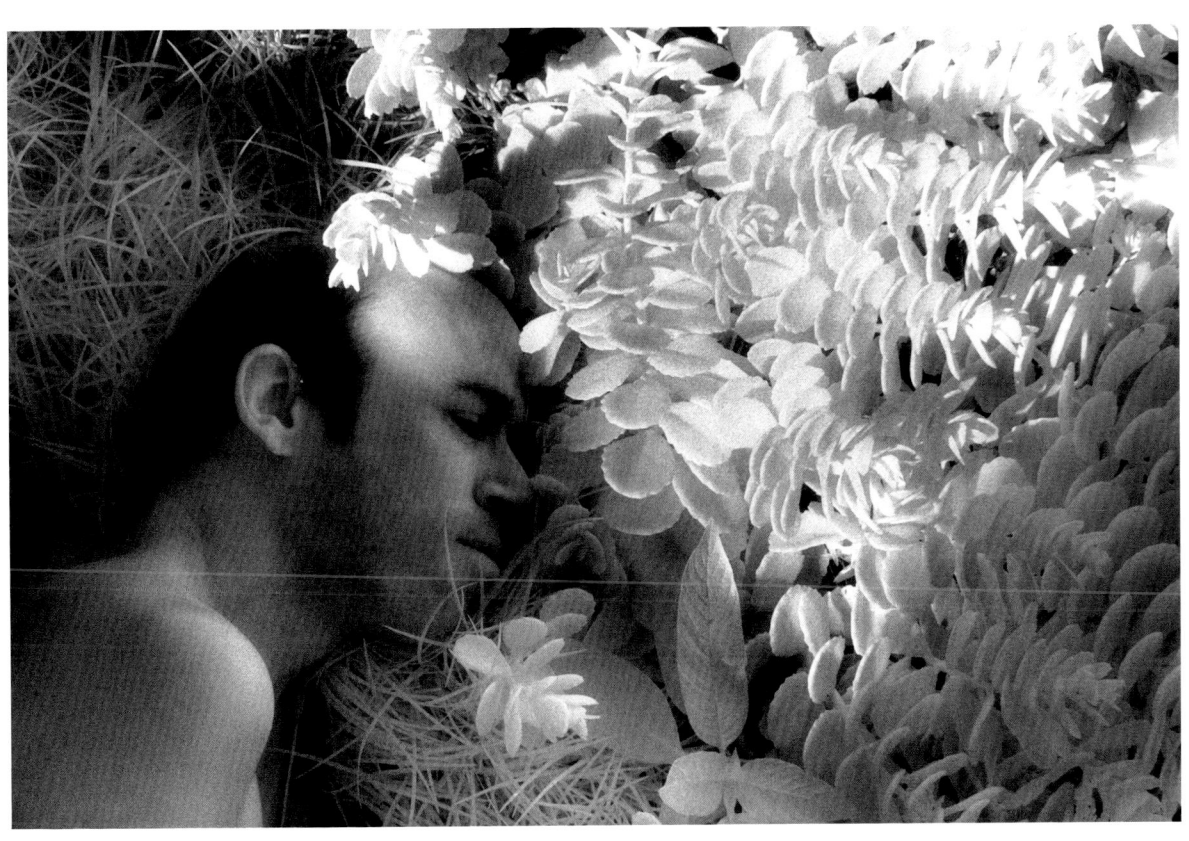

ANNIE ROGERS
Honolulu, Hawaii
Garden Dreams

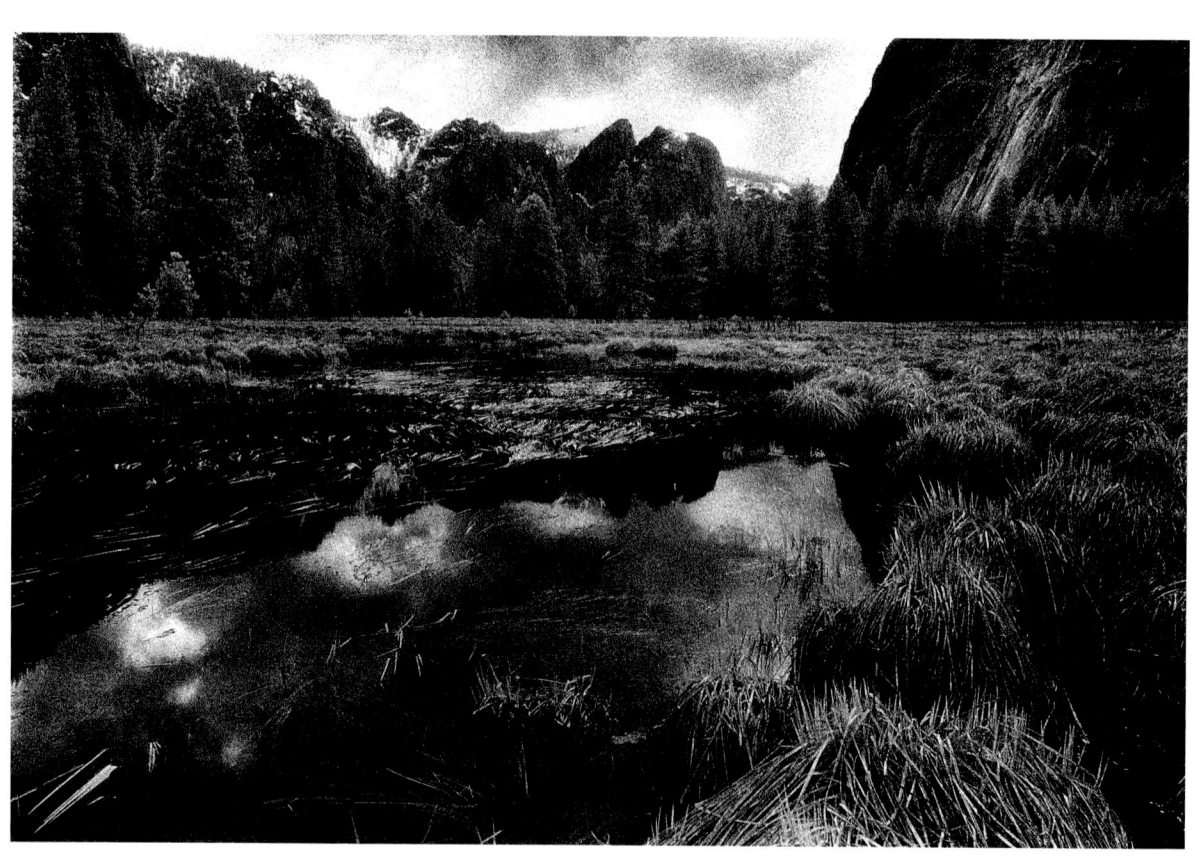

JIM BENEDICT
Moose, Wyoming

Yosemite Valley, Yosemite National Park, California, 1976
(©Jim Benedict, 1982)

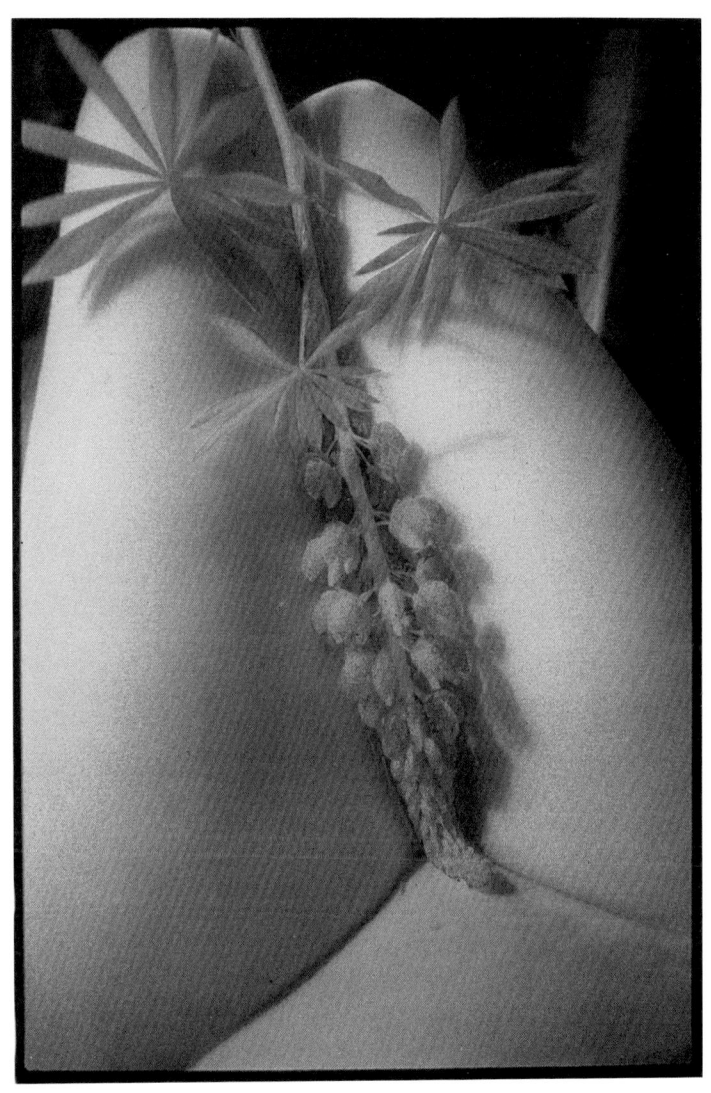

ANNE KURUTZ
Rockport, Maine
Lupine (hand-colored with airbrush and pastels)

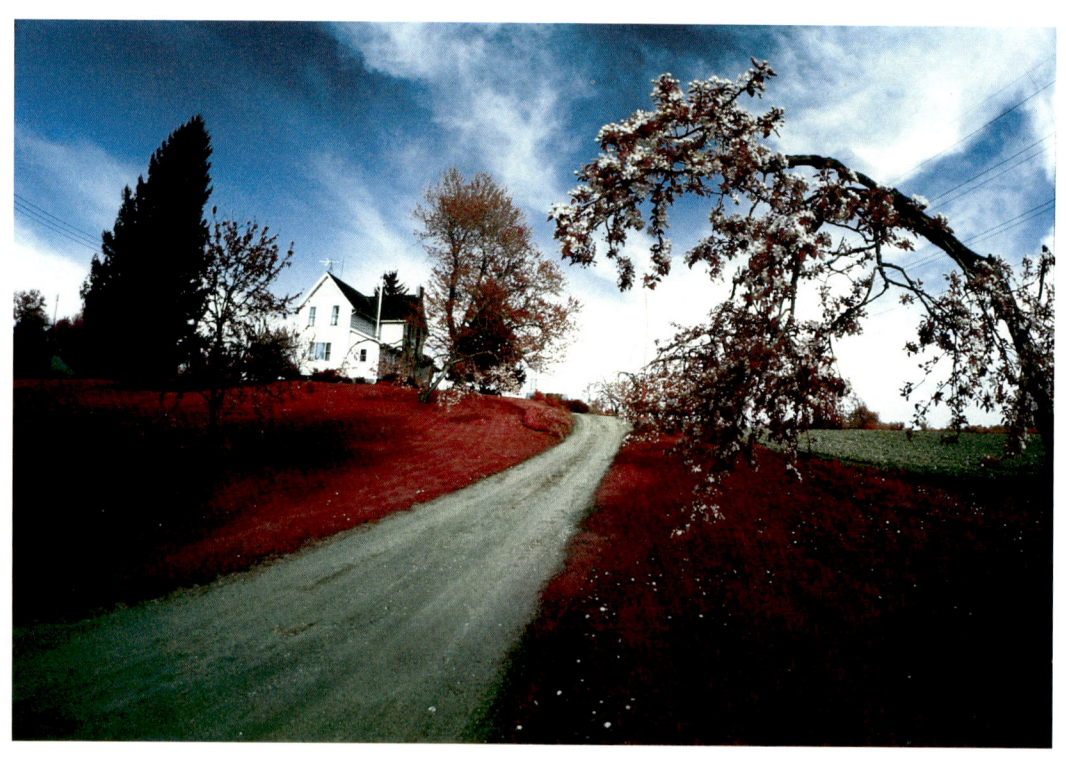

DAVID PATERNITE
Akron, Ohio
Untitled, 1982

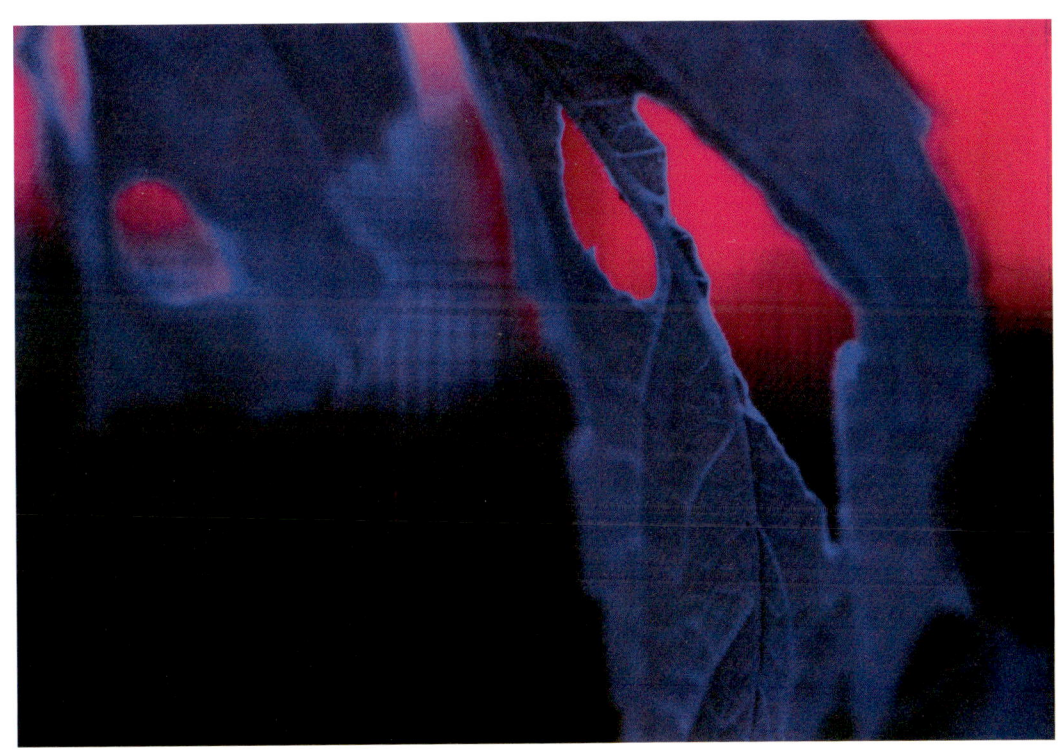

HANS ROPERS
North Canton, Ohio
Untitled

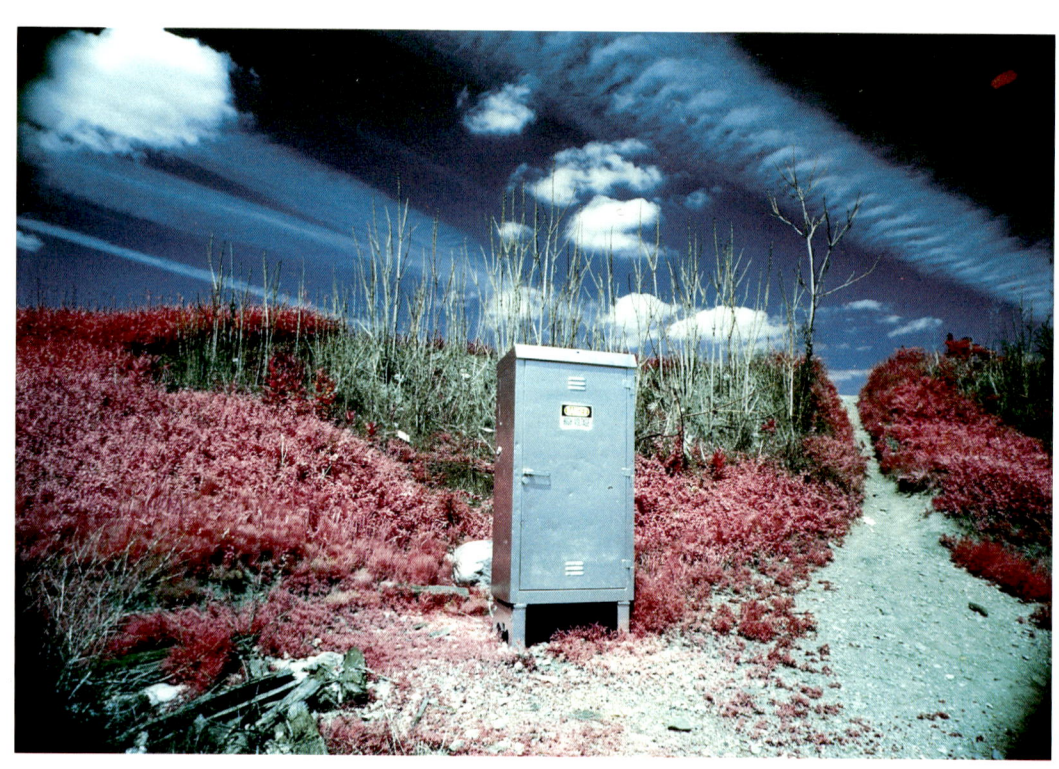

STEPHEN PATERNITE
Akron, Ohio
Untitled, 1982

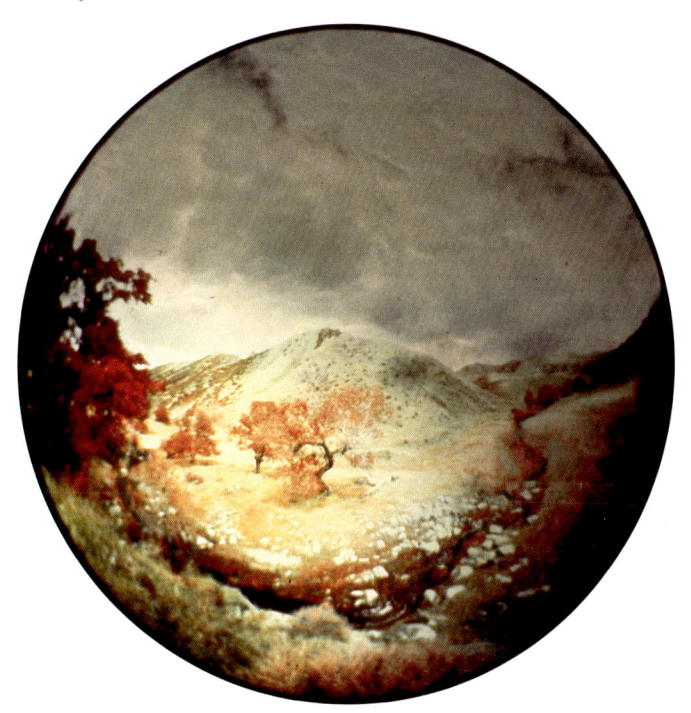

MARY WALSH-LAHTI
Volcano, Hawaii
First Meditation (sandwiched emulsions)

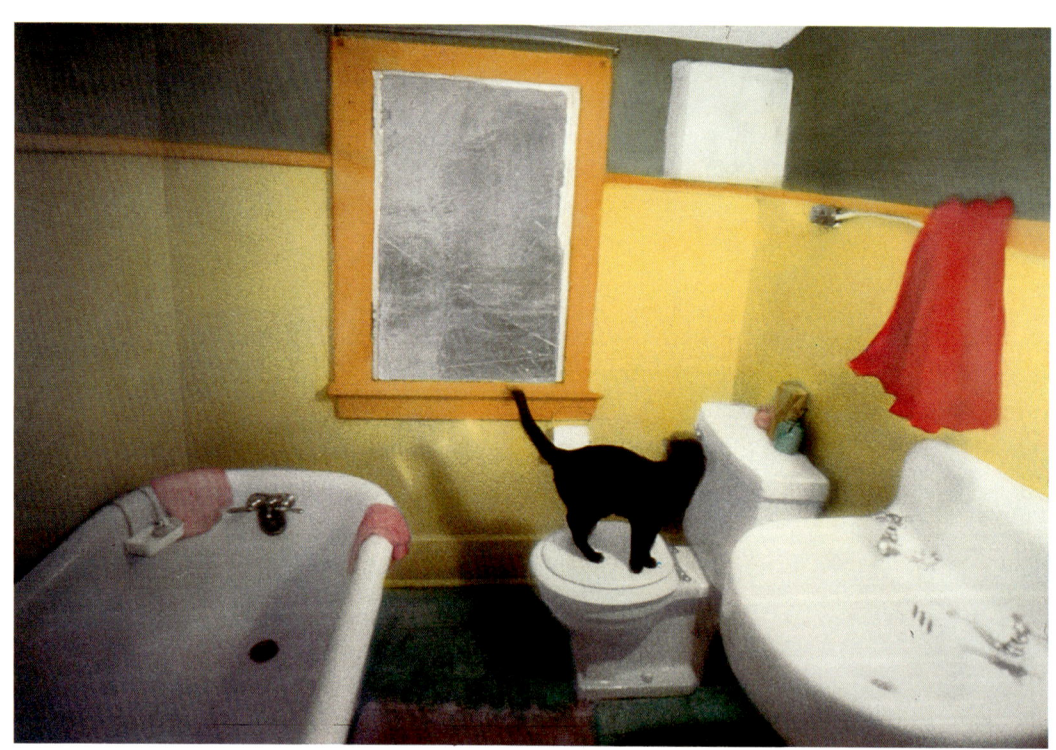

RITA DIBERT
Mt. Baldy, California

Fernie's Bathroom, Columbia, Missouri, 1978 (tinted 1981)
(hand-tinted b&w photograph-original, 4' x 6')
courtesy: Xochipilli Gallery, Birmingham, MI

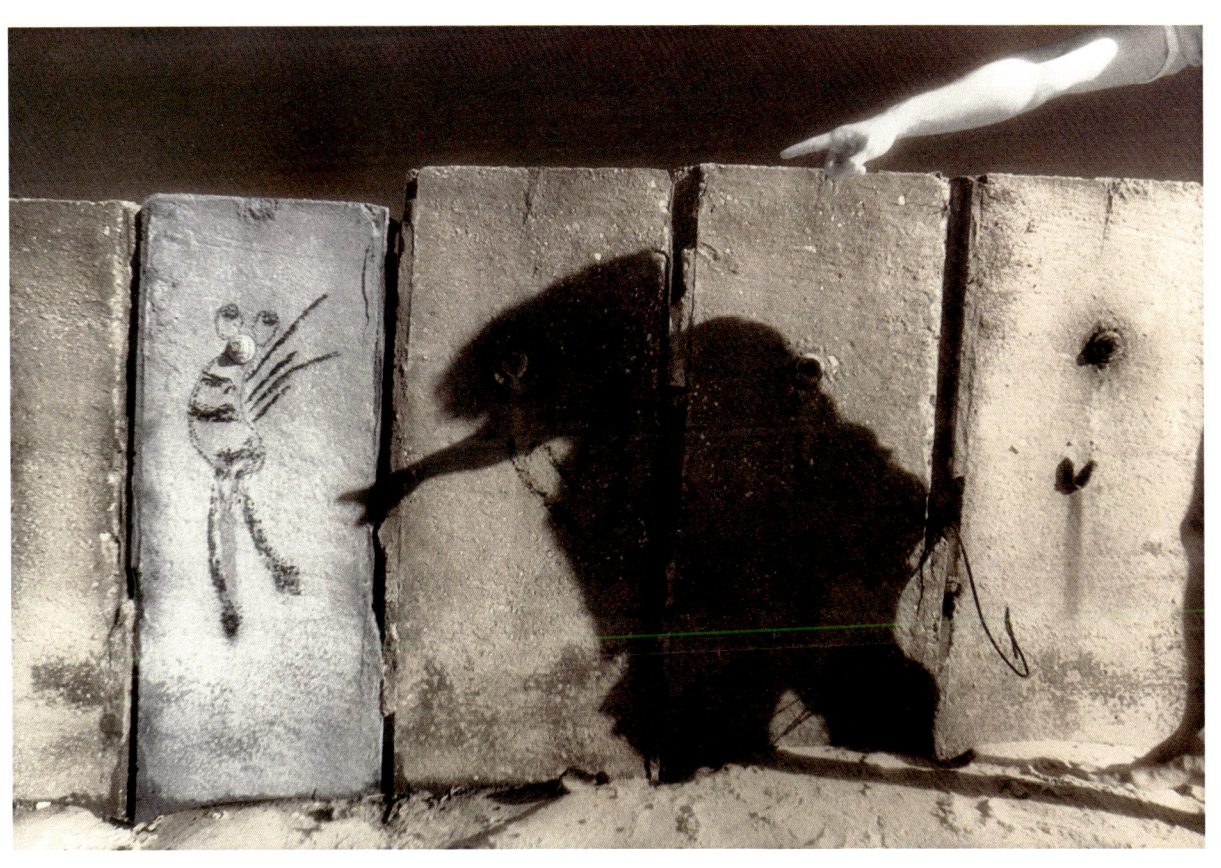

BARBARA EDWARDS
Tallahassee, Florida
Man And Creature

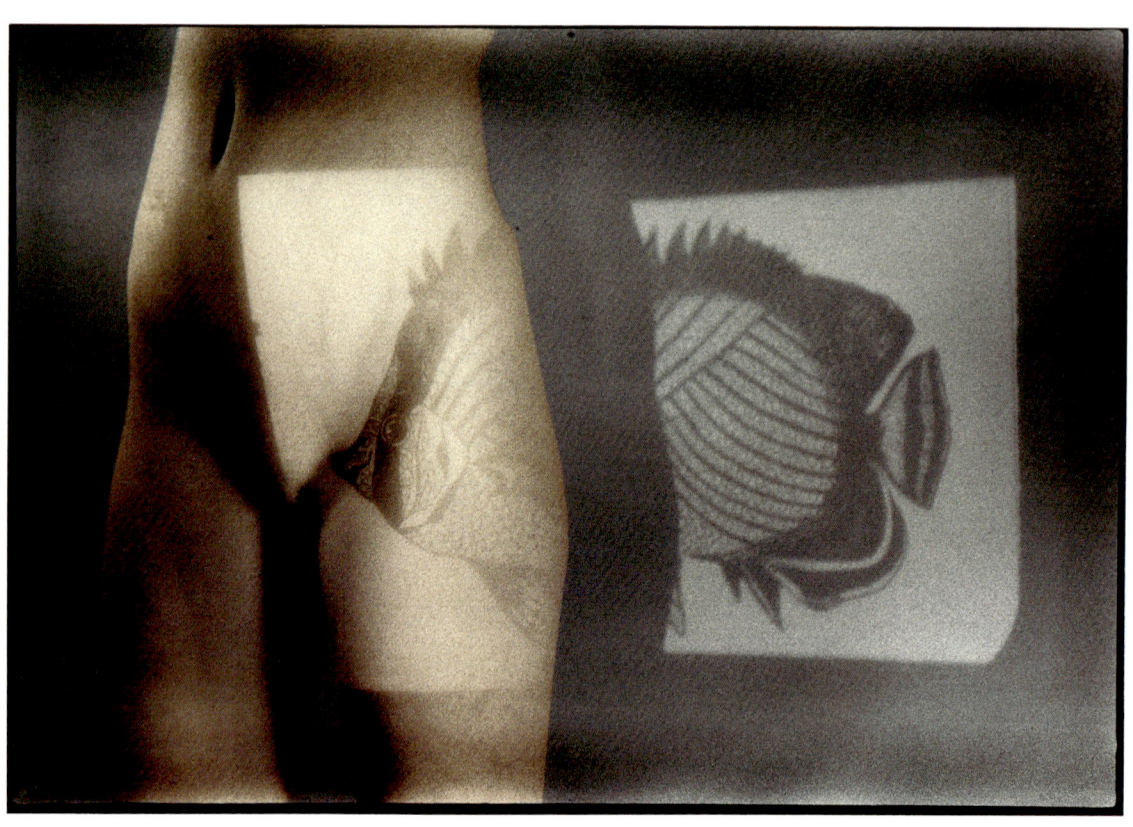

DENNIS LETBETTER
San Francisco, California

"Tropical Fish" from the "Projected Nude" series

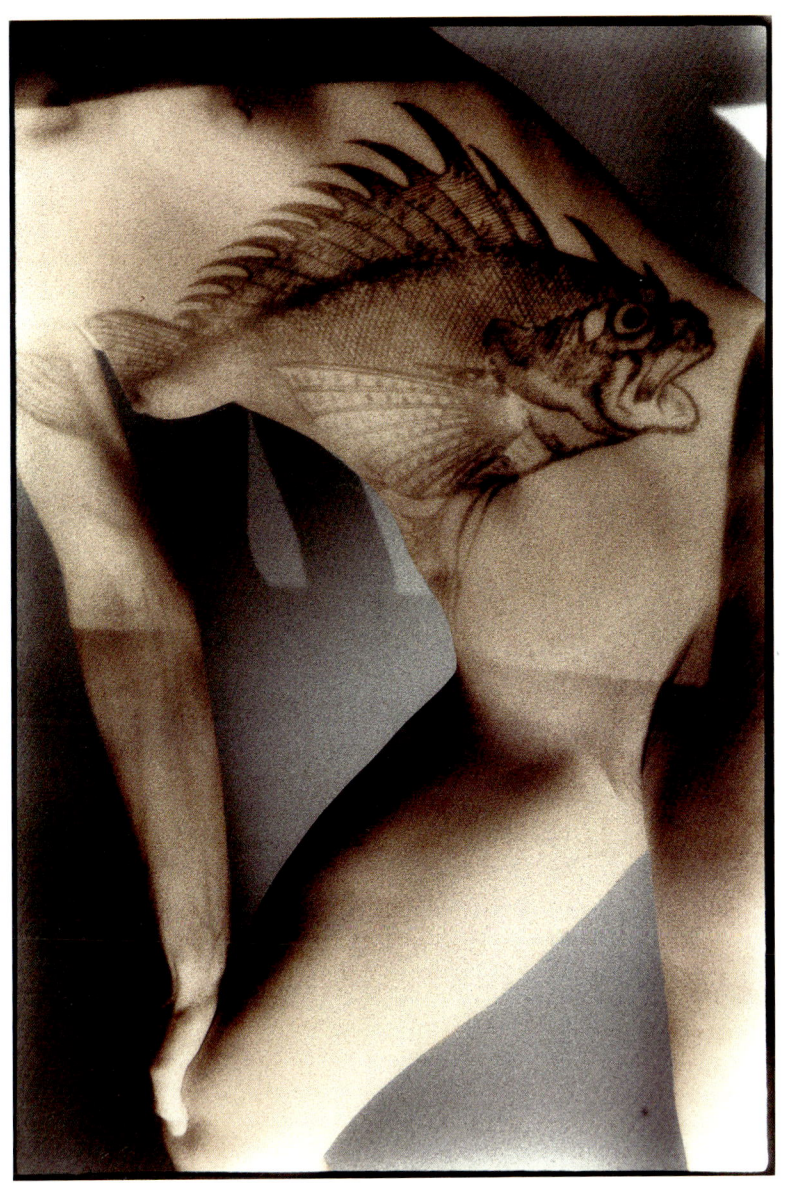

DENNIS LETBETTER
San Francisco, California
"Fish #3" from the "Projected Nude" series

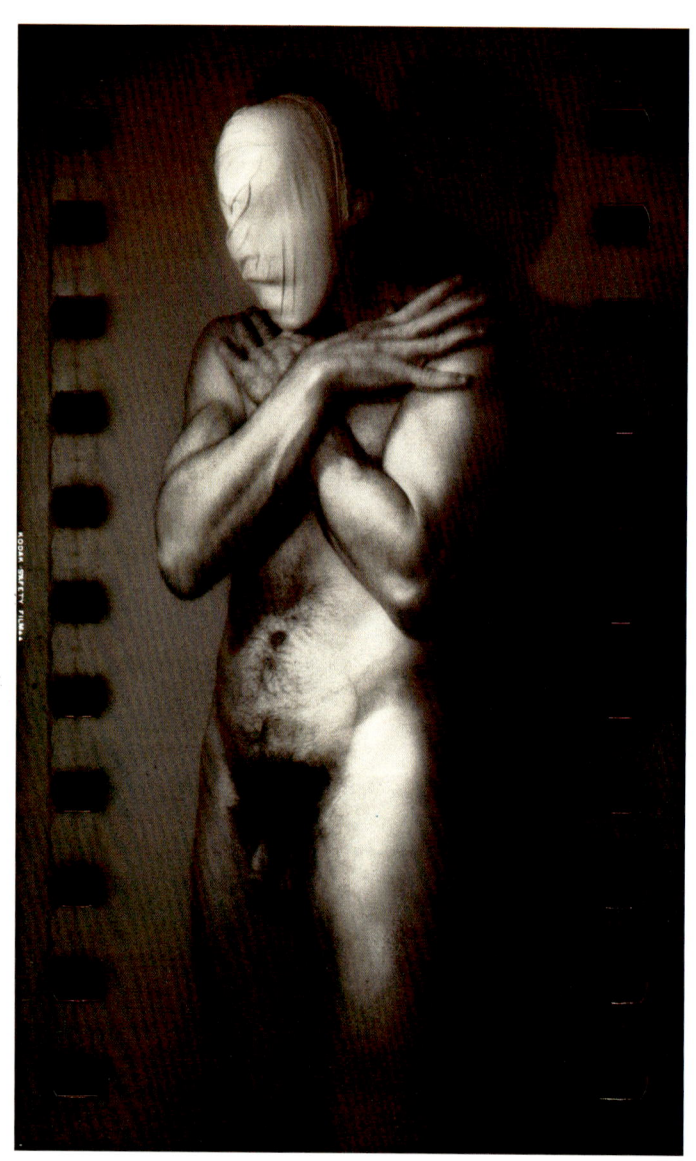

JOHN HOOPER, JR.
San Antonio, Texas
Untitled

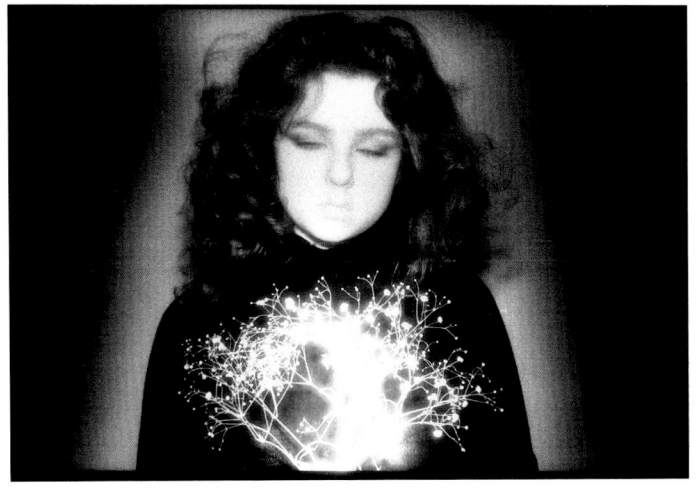

SHIU LEUNG
Nashville, Tennessee
Untitled

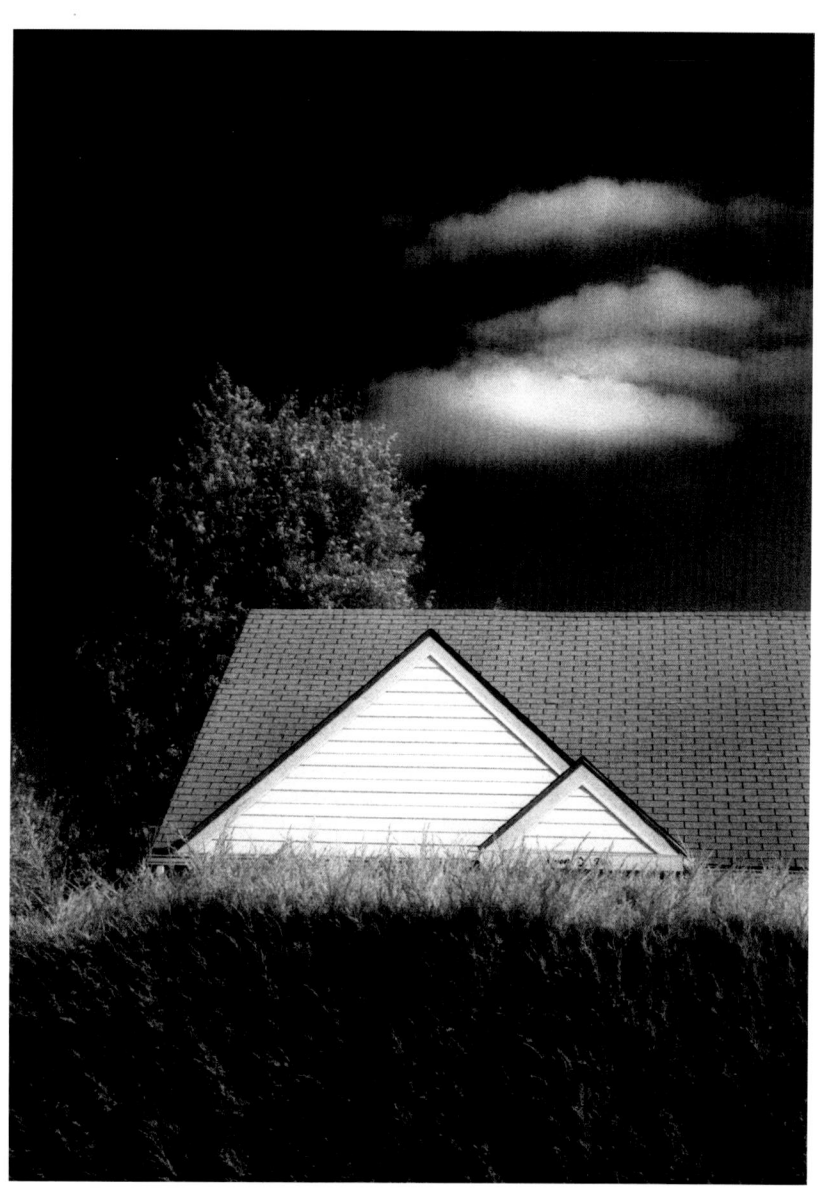

THOMAS H. SHULER, JR.
Washington, D.C.
Untitled

THOMAS H. SHULER, JR.
Washington, D.C.
Untitled

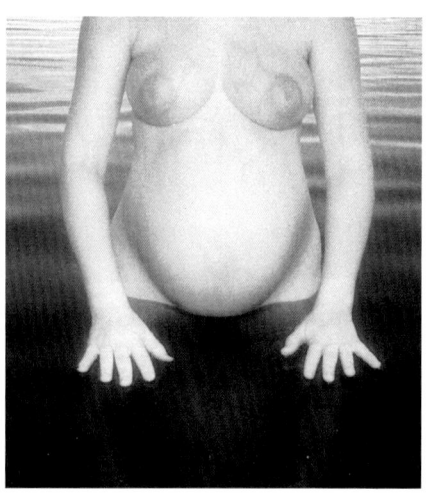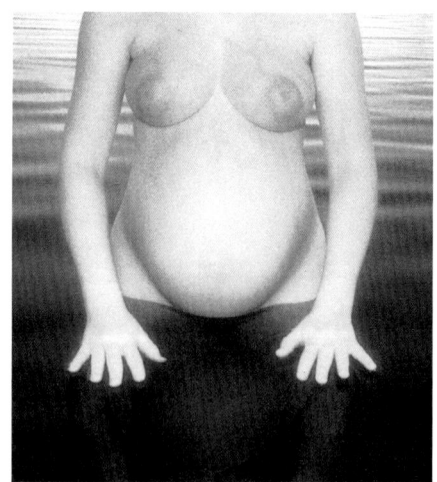

STEVEN SCHWARTZMAN
Austin, Texas
Untitled From The "Pregnancy Series" 1979 (stereo)

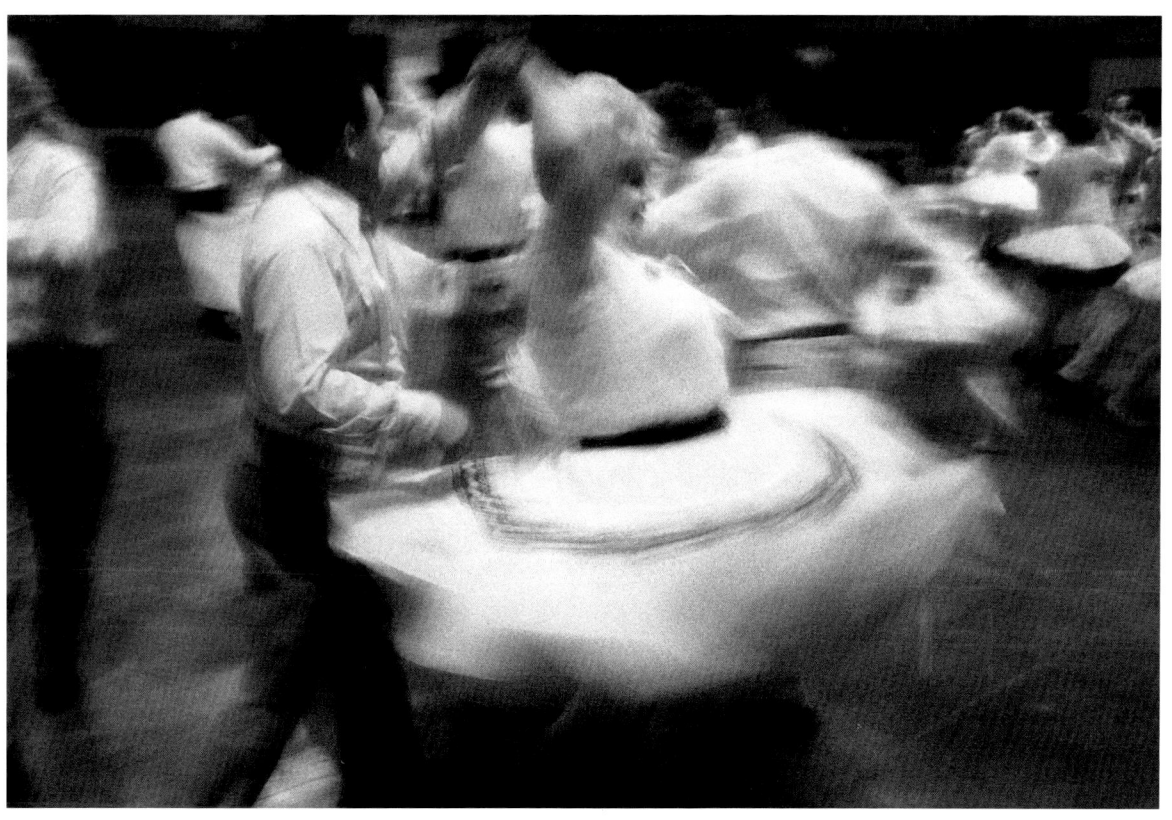

CHRISTOPHER BURNETT
Albuquerque, New Mexico
"Swirling Dancers" from the "Pearl Promenaders" series

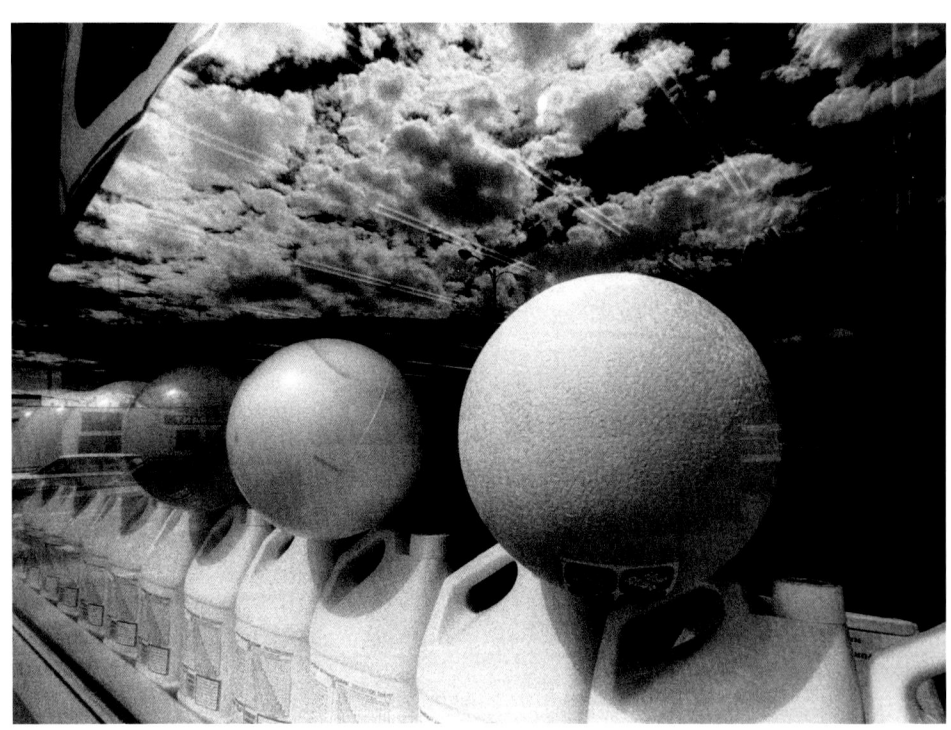

ANNIK BRUNET
New York, New York

Supermarket Window, Eureka, California, 1978

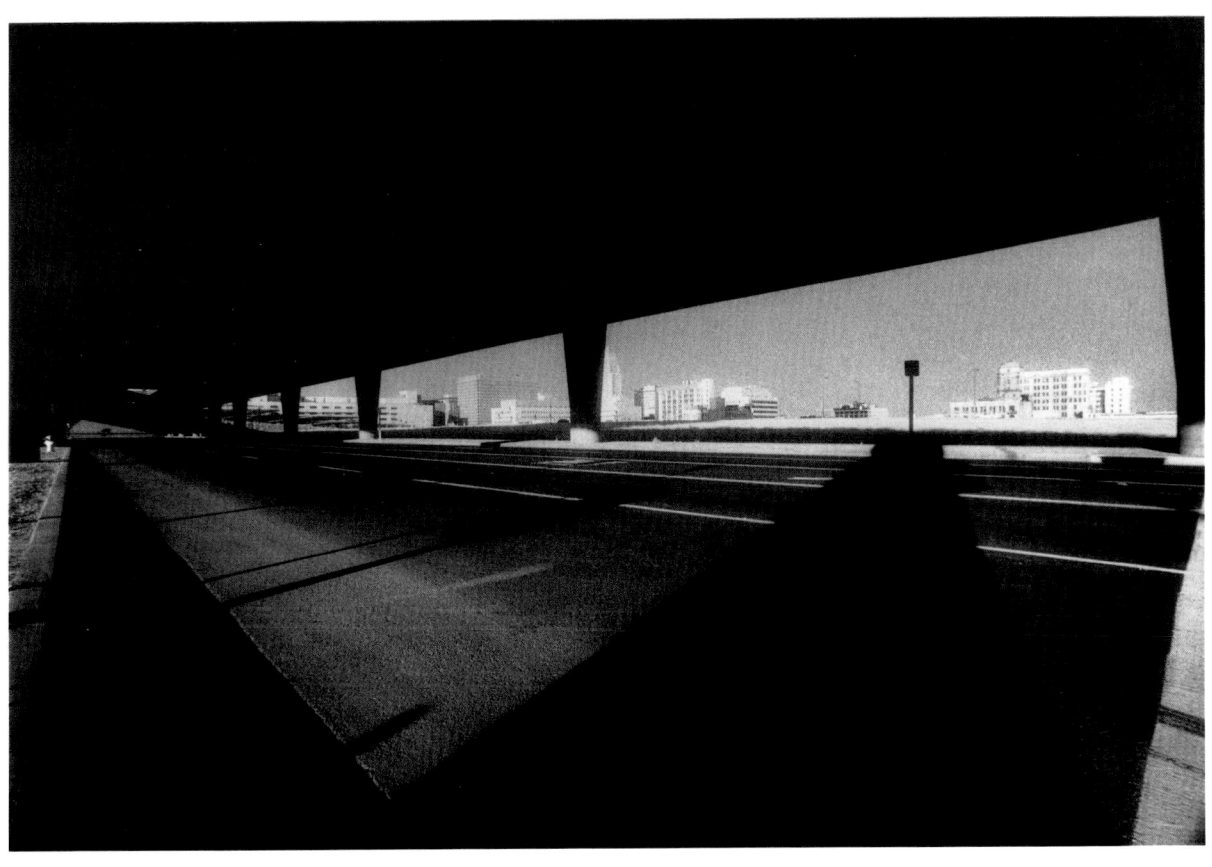

ANNIK BRUNET
New York, New York
Downtown Freeway, Los Angeles, California, 1976

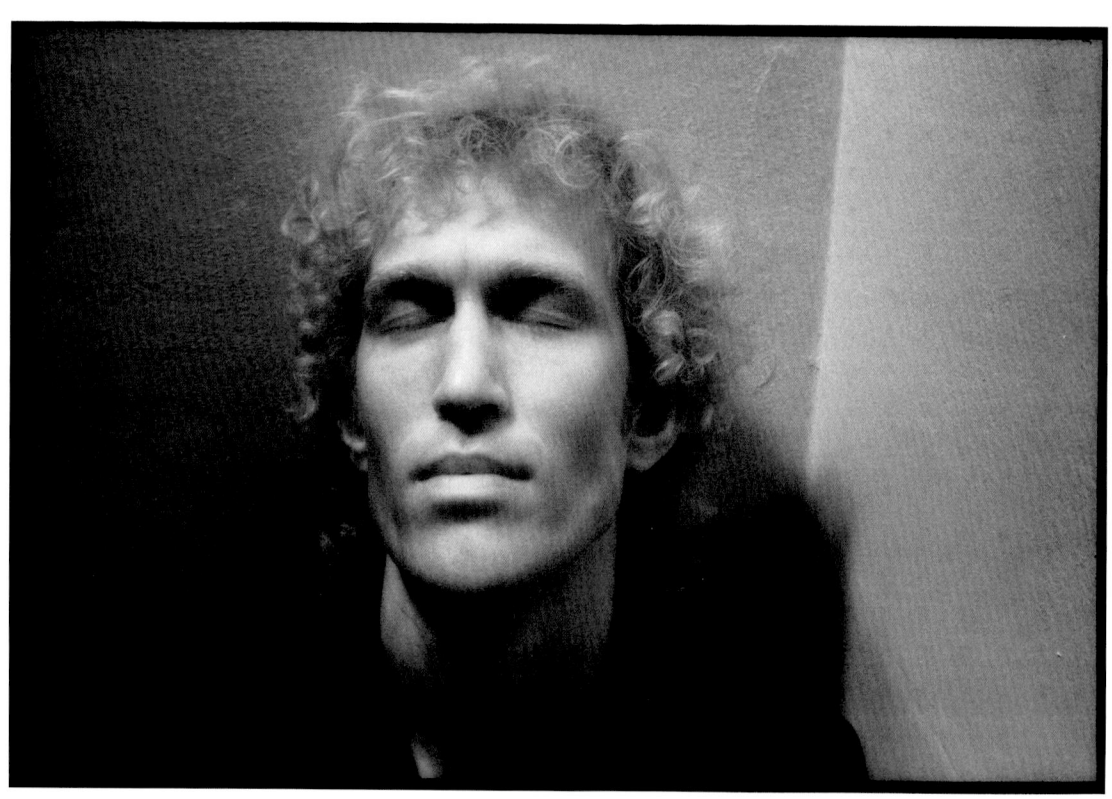

MAURA ROBINSON
Richmond, Utah
Portrait Of Dee, 1982

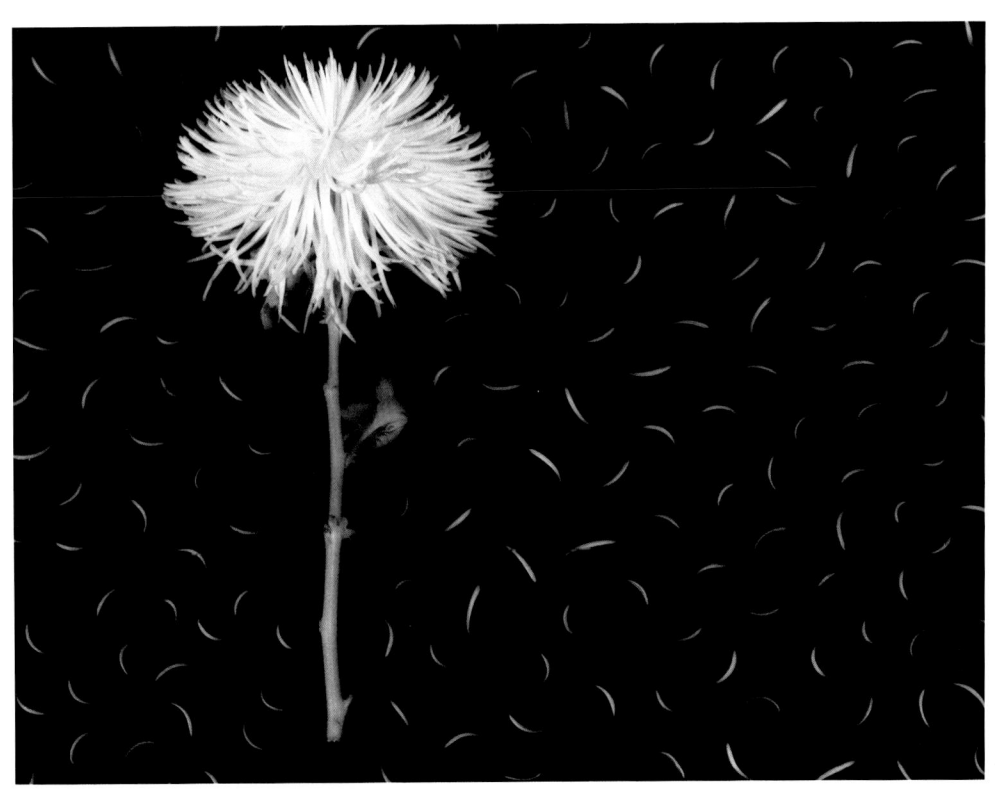

GAIL LeBOFF
New York, New York
Untitled

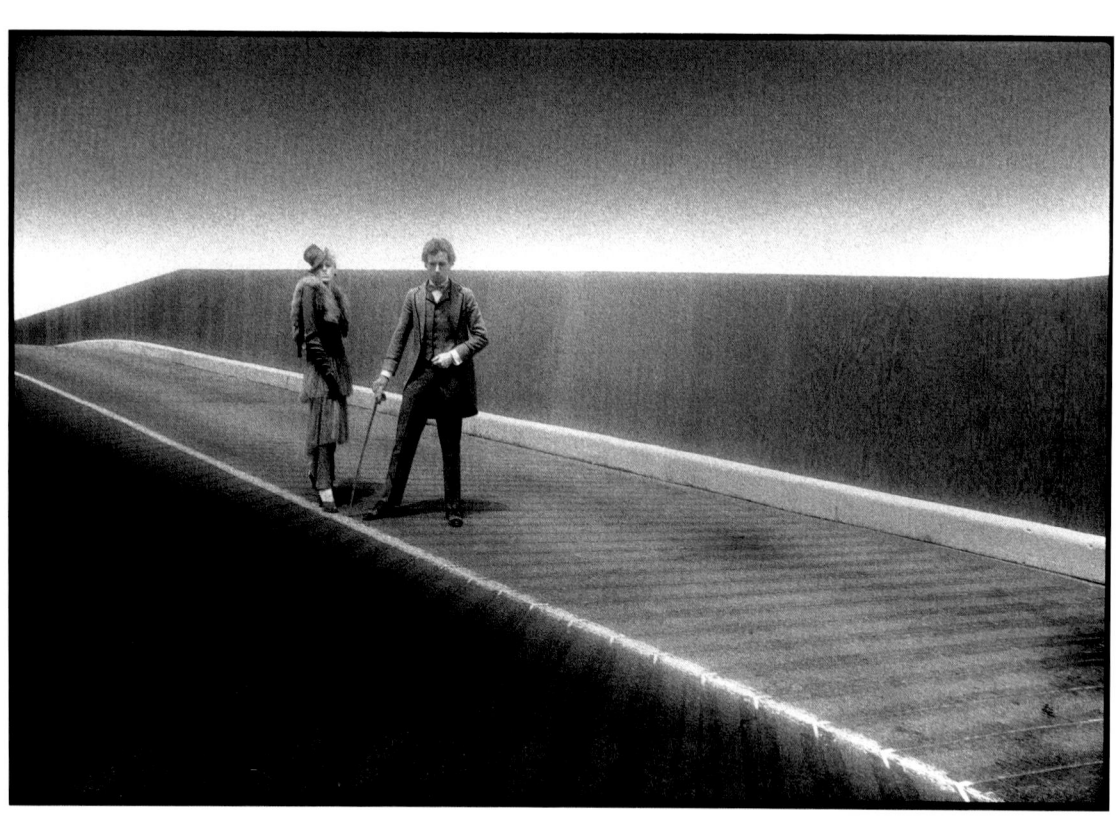

LIN EAGLE
Chicago, Illinois
Untitled

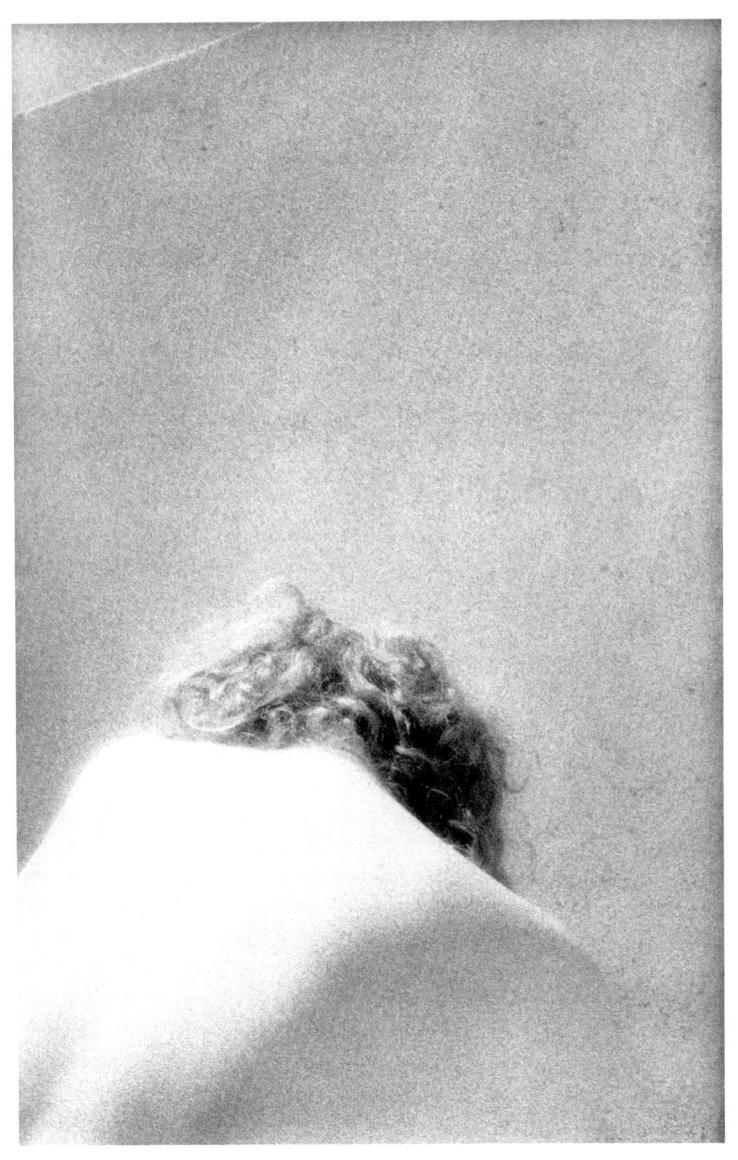

PAMELA J. LANDAU
Hoboken, New Jersey
The Back

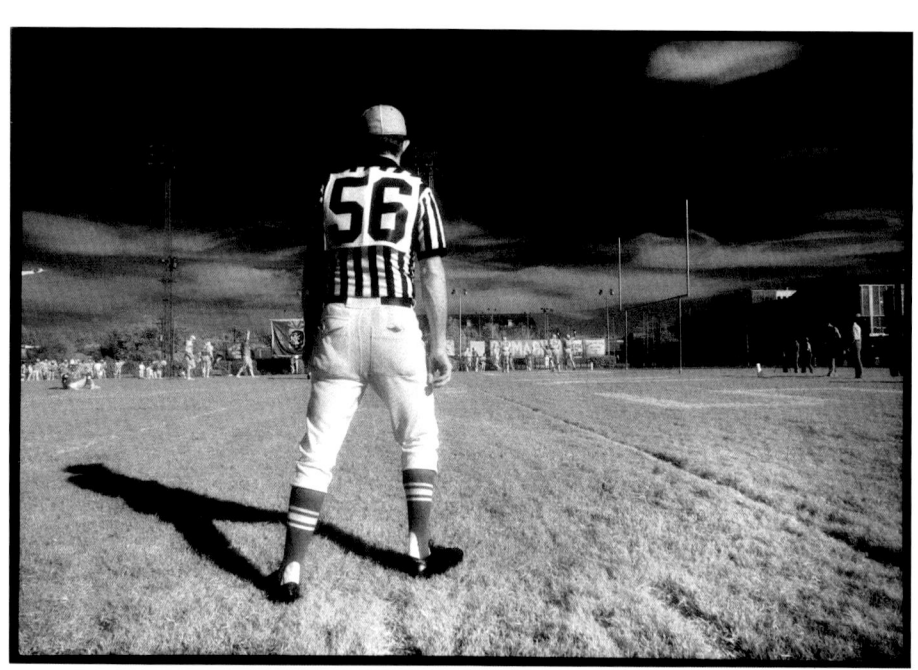

BRIAN COATS
Commerce, Texas
Fifty-Six

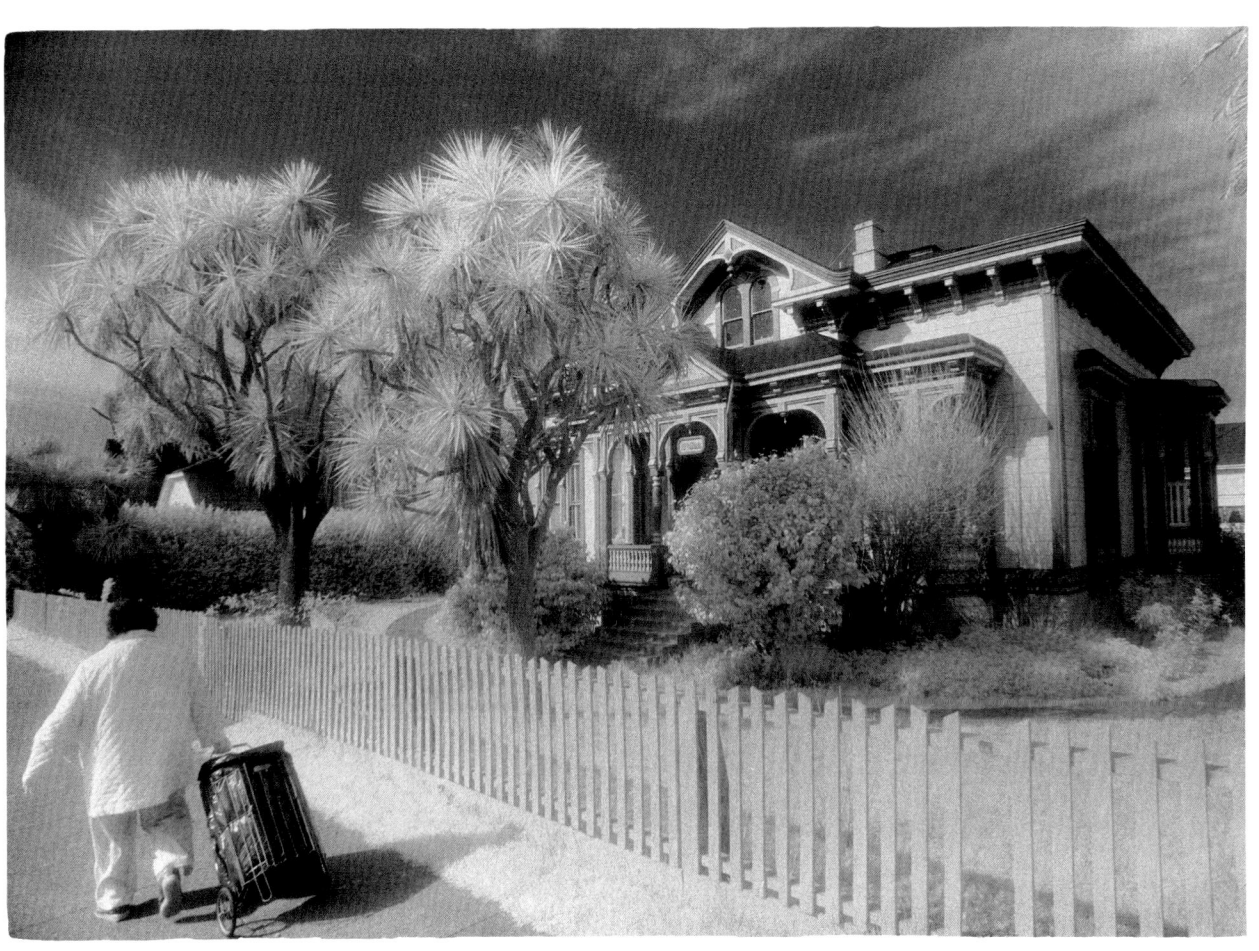

NEVILLE GODFREY
Arcata, California
Untitled

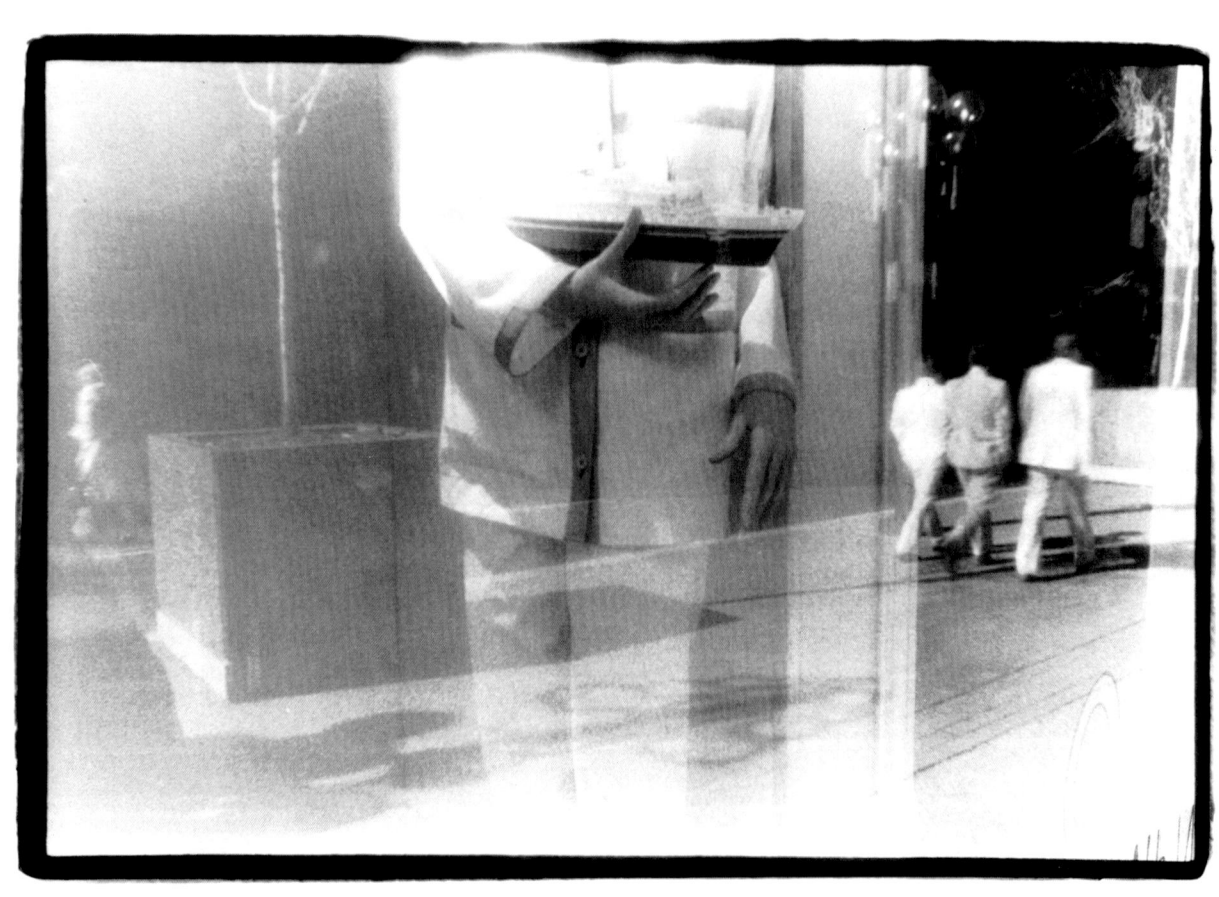

STEPHEN SPERA
Philadelphia, Pennsylvania
Untitled

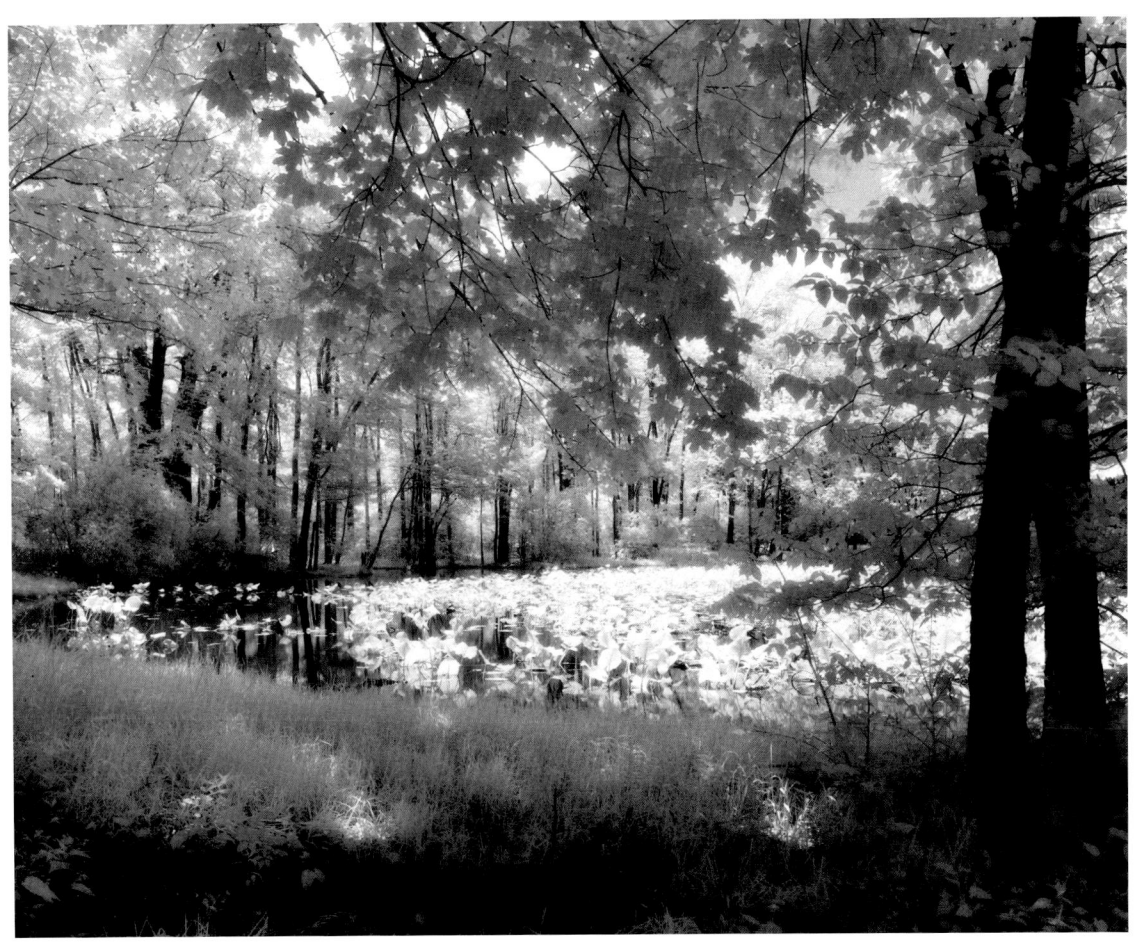

RANDY JUSTER
Bogata, New Jersey
Pond, Florham Park, New Jersey, 1978

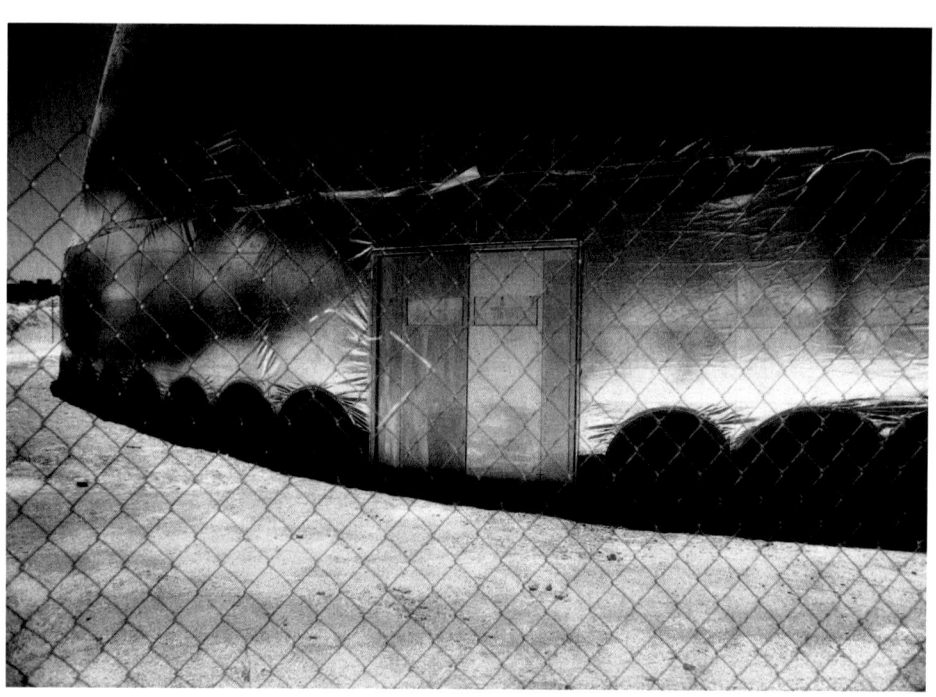

CAROL LACHATA
Pasadena, California
Untitled

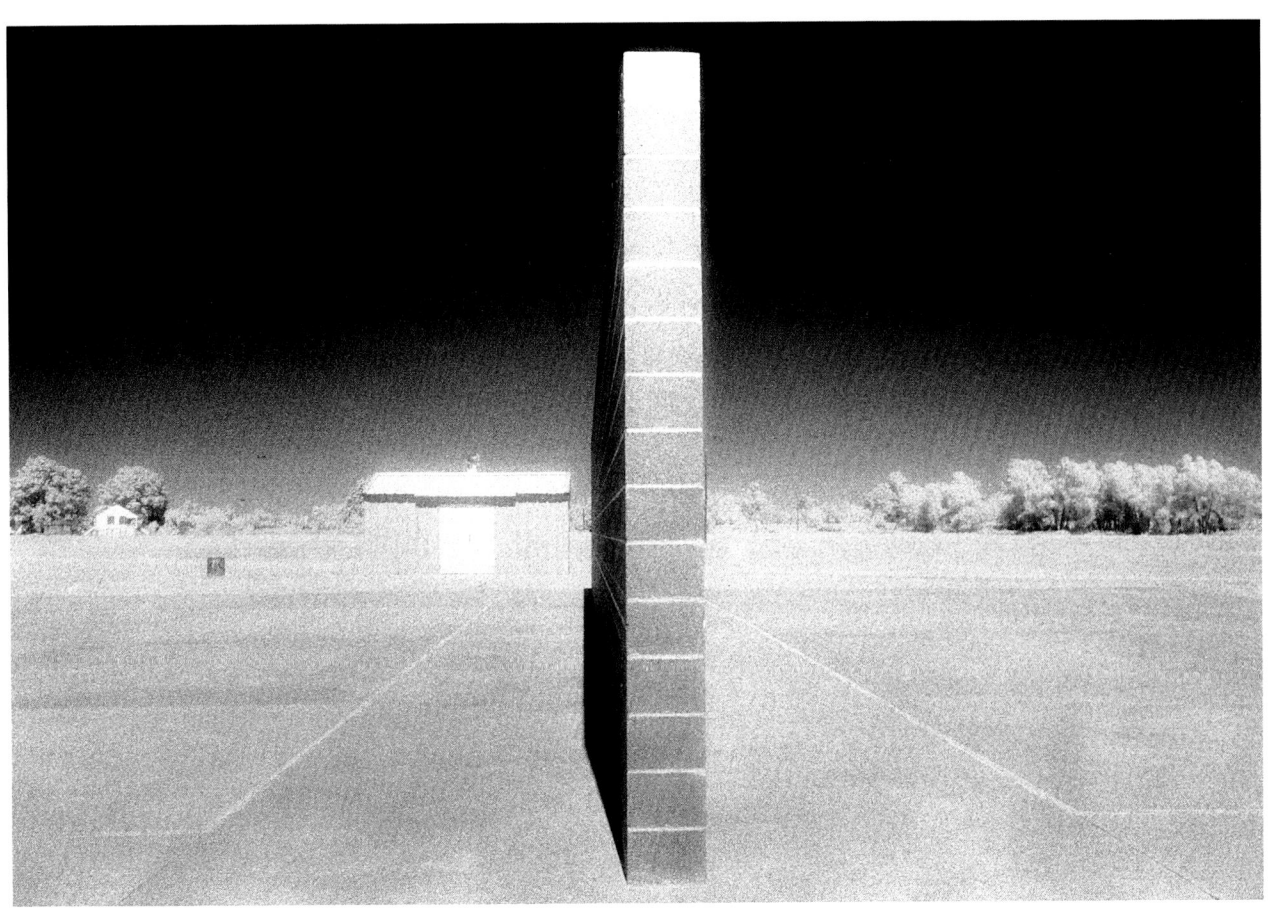

GARY CAWOOD
Rustin, Louisiana
W. Eugene Smith Monument, 1978

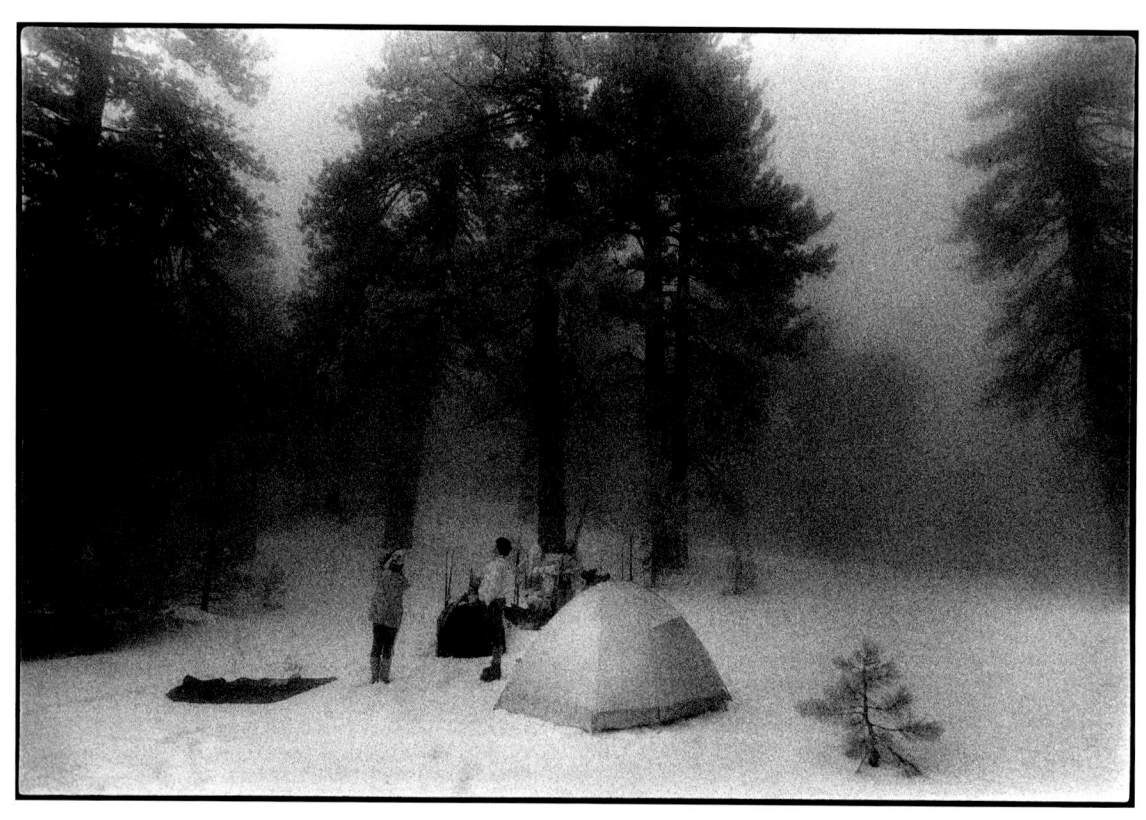

CHARLIE MARTIN
Los Angeles, California
Snow Camping

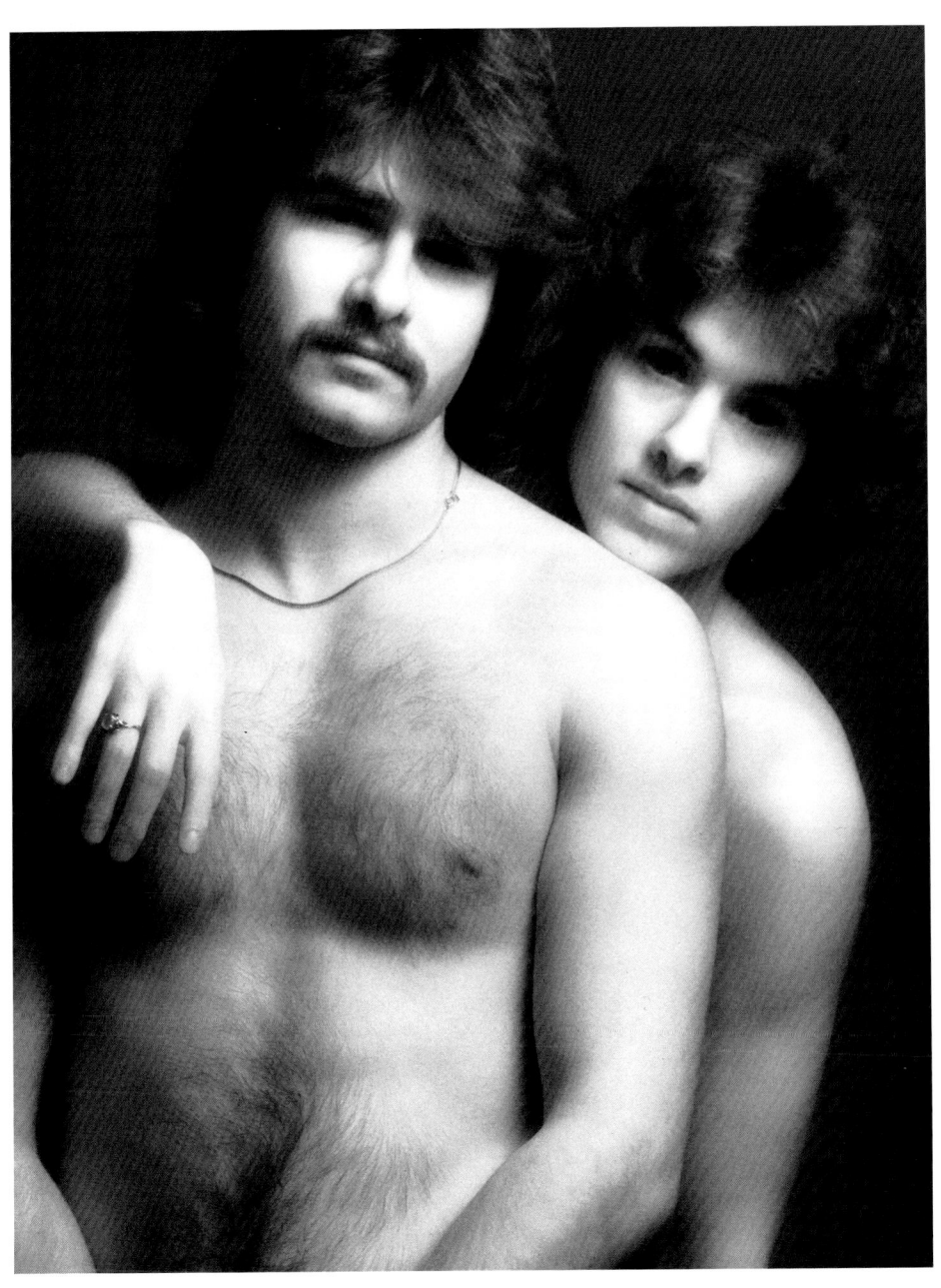

GARY GANTERT
Jackson, Michigan
Self Portrait With Sister

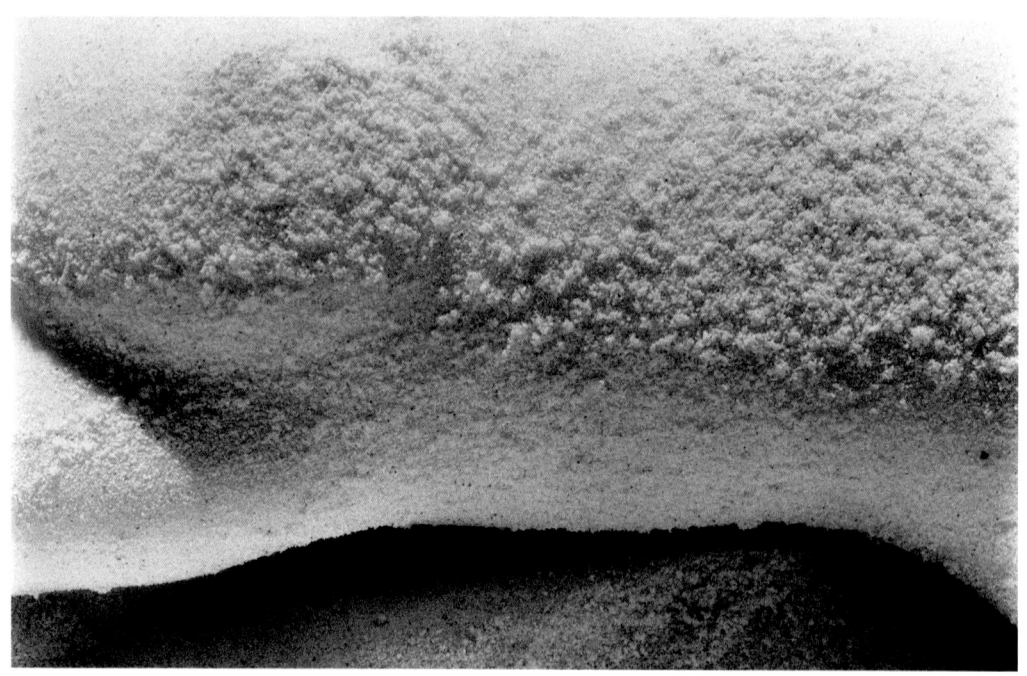

GIORGIO MAJNO
Carbondale, Illinois

Amona, 1982

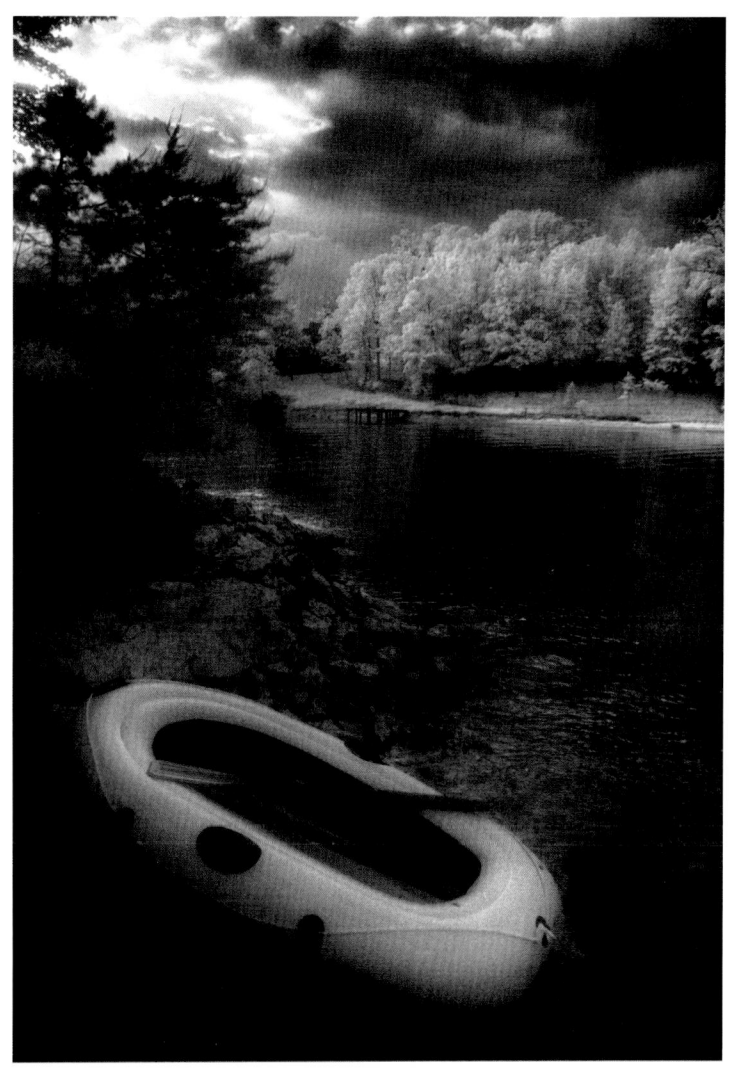

EDWARD WADDELL
Charlotte, North Carolina
Lakescape #2

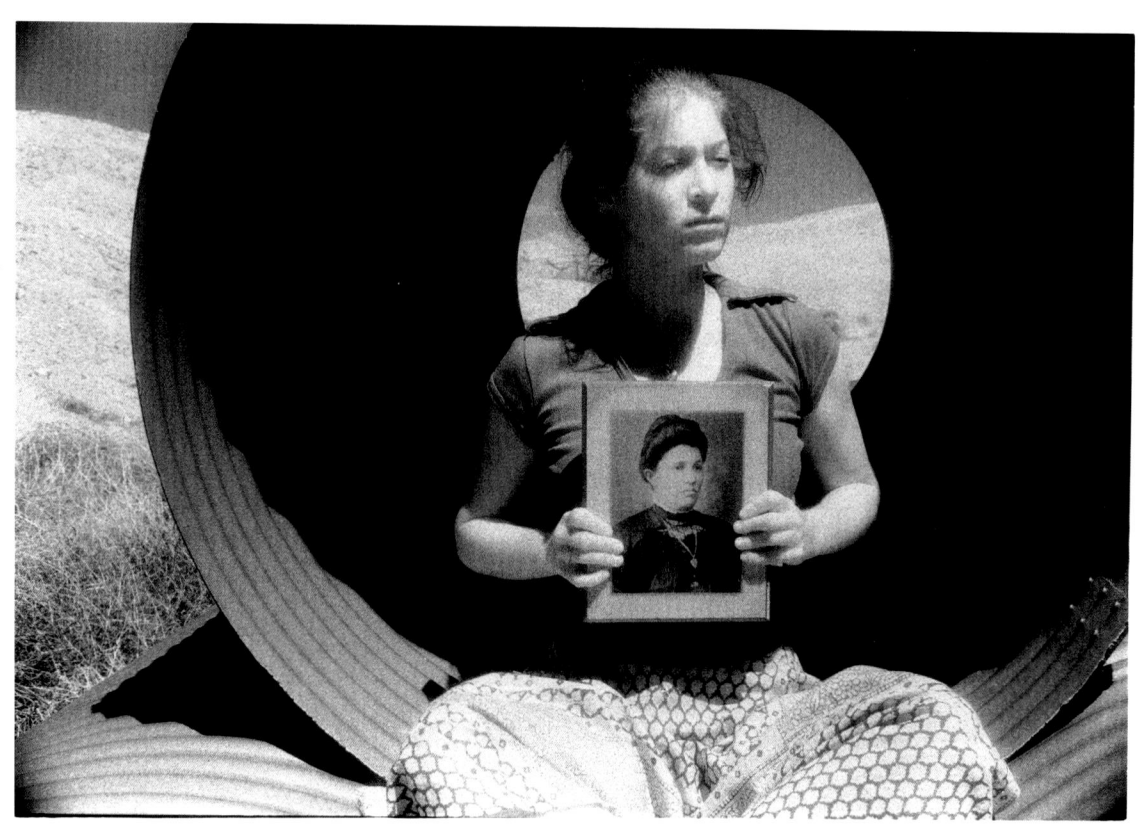

LELA HERSH
Urbana, Illinois

My Mother's, Mother's, Mother

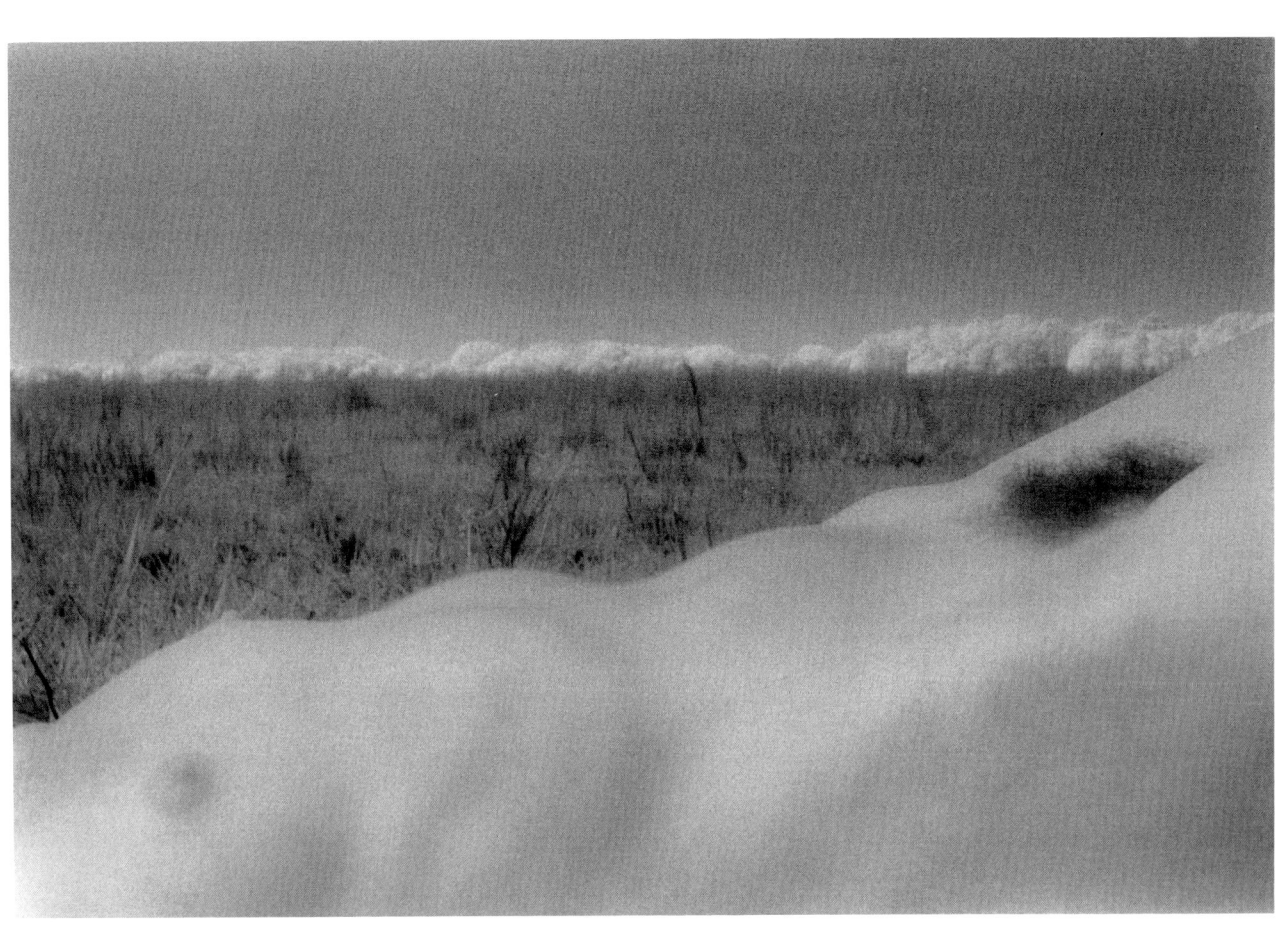

SUSAN J. MOORE
Daytona Beach, Florida
Untitled

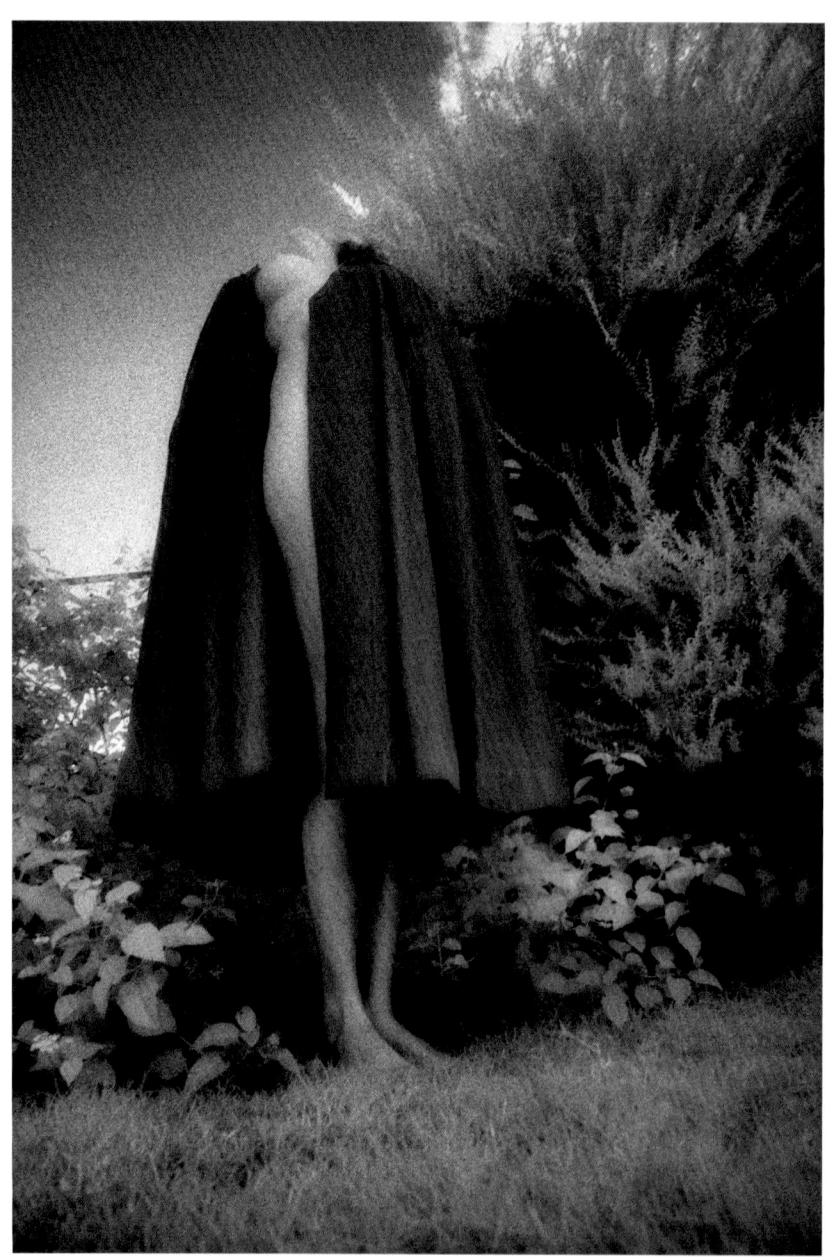

GAVIN MORRISON
Chicago, Illinois
Culmination, 1975

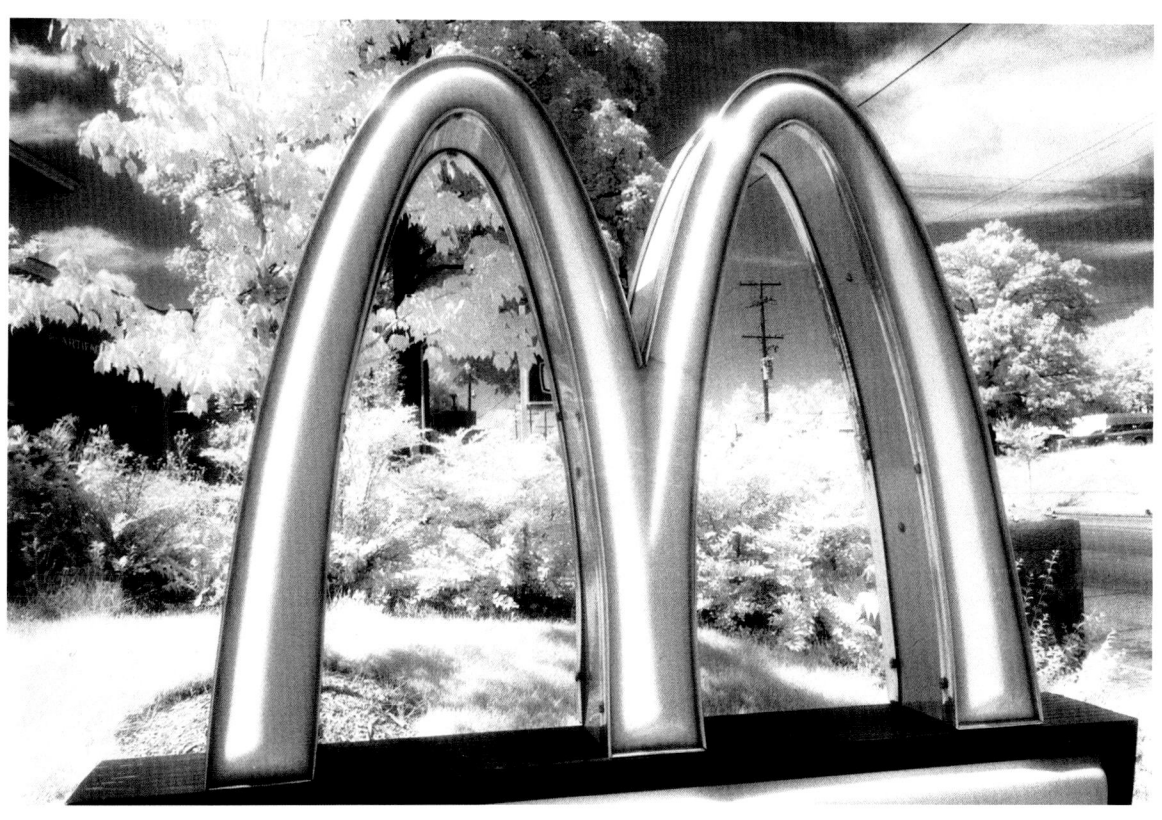

CHRISTOPHER GALLAGHER
Columbus, Ohio
Untitled

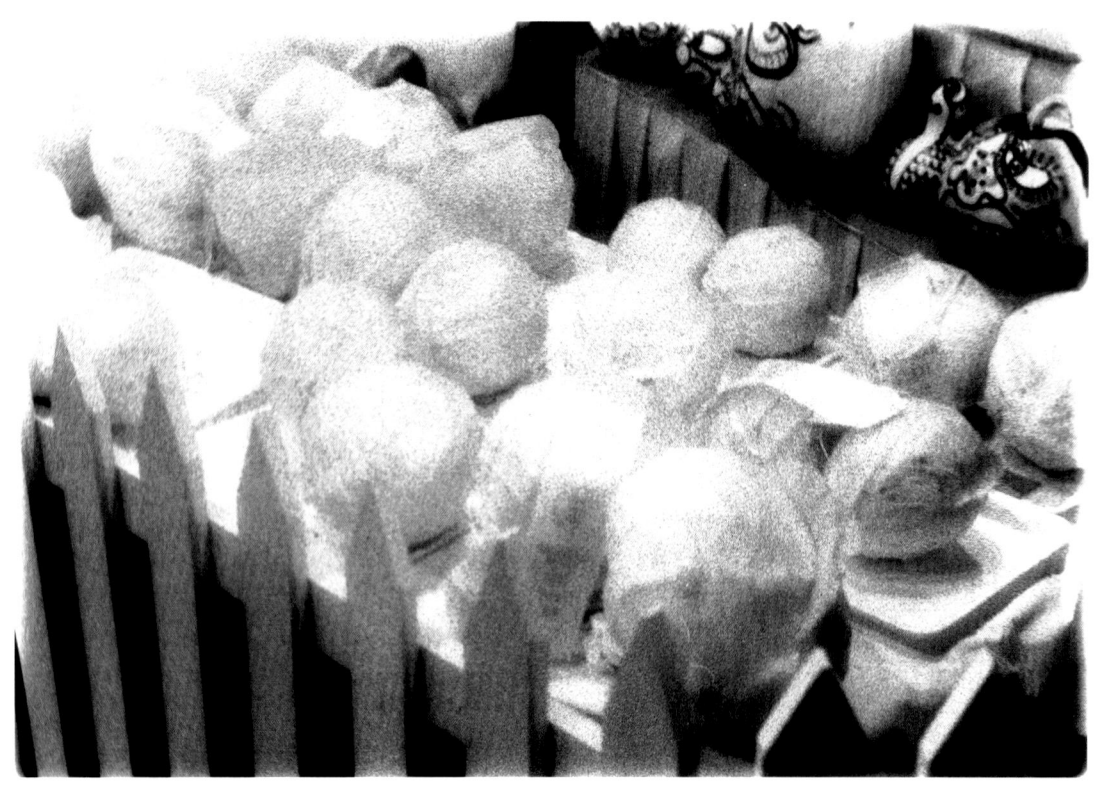

JEROME MILLER
Detroit, Michigan
State Fair #*3*-Cabbages

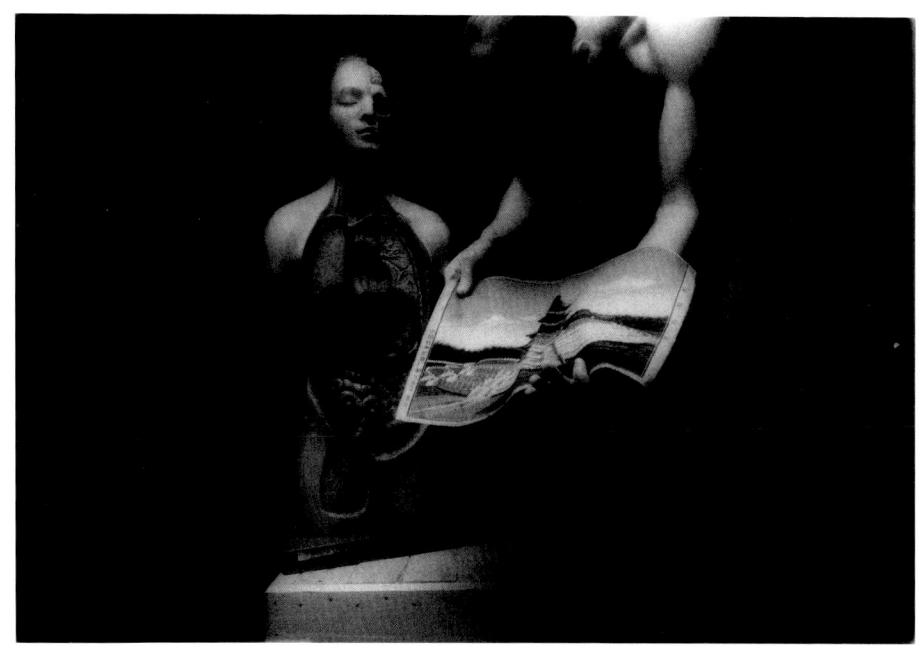

PETER IVERSON
Tallahassee, Florida
I Read the News Today, #12

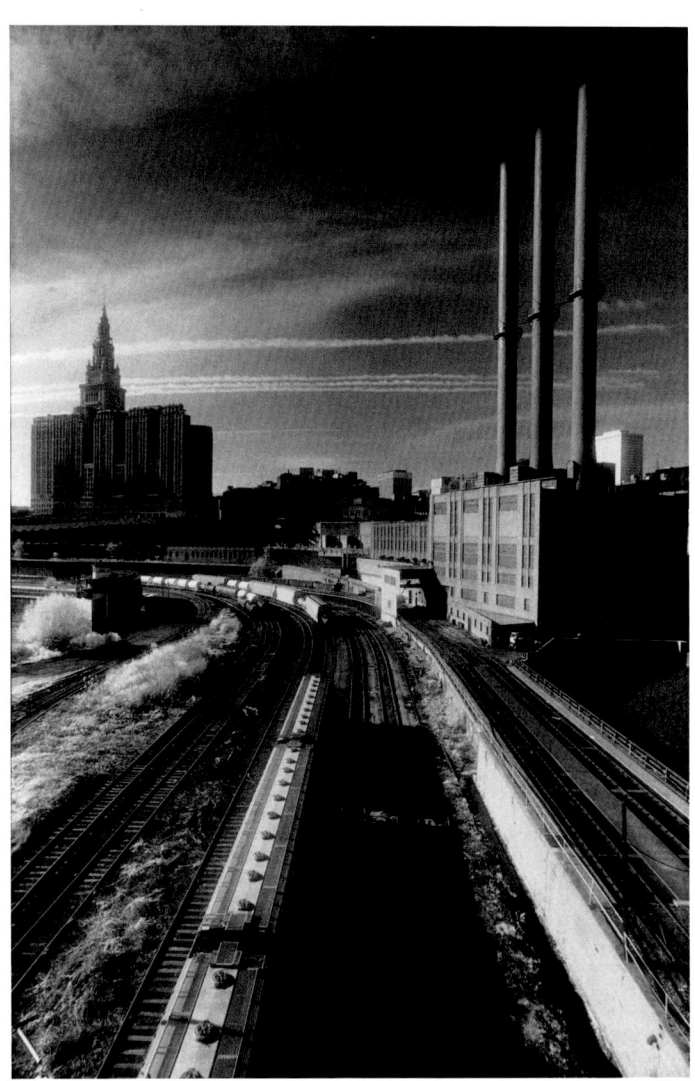

ABE FRAJNDLICH
Cleveland, Ohio

Rail Tracks and Terminal (© Abe Frajndlich, 1981)
(from the book *Cleveland Infra Red,* Publix Imprints)

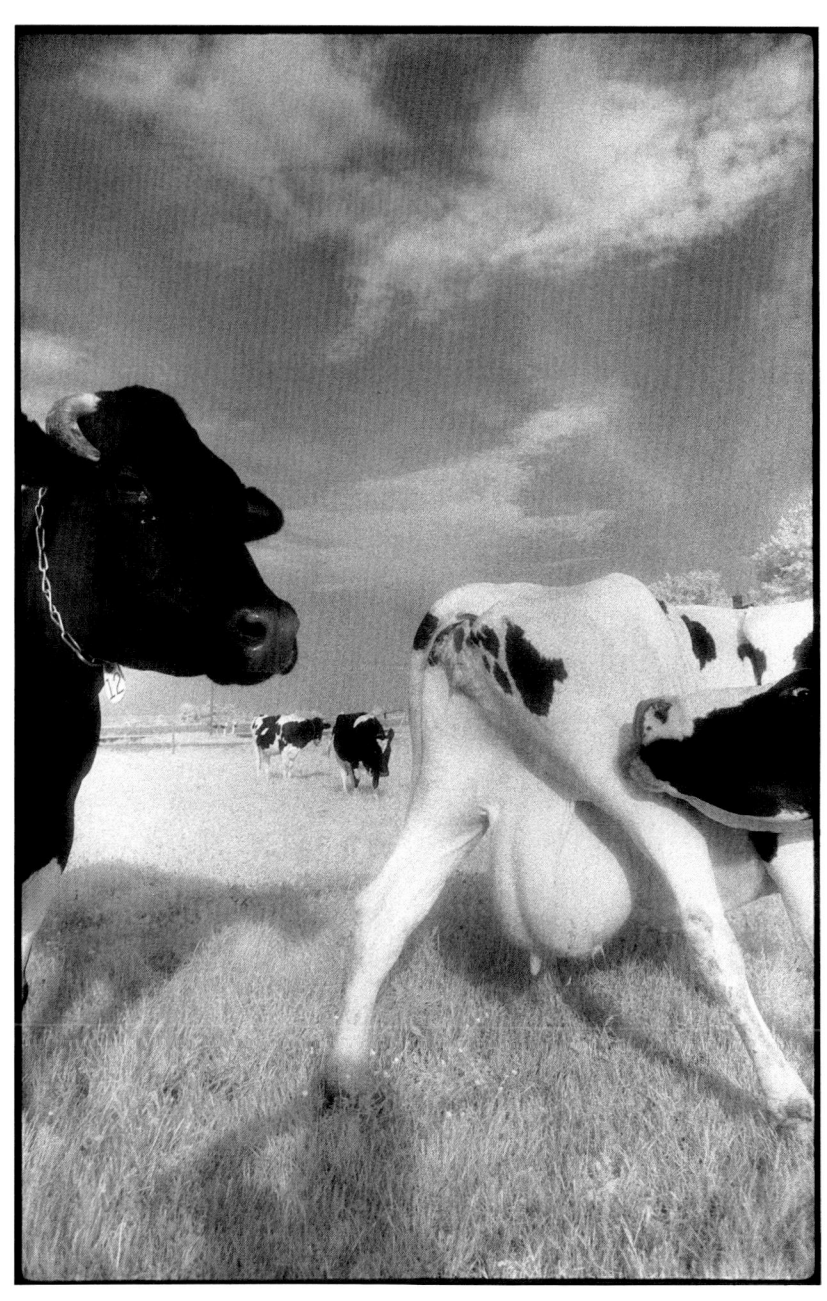

PENNY RAKOFF
Akron, Ohio
Untitled

THE PHOTOGRAPHERS

William D. Adams, Houston, TX
Michael Addobati, Sacramento, CA
Richard Albright, Pocatello, ID
Jim Alley, Interlochen, MI
James J. C. Andrews, New York, NY
Linda Anthis, Twin Falls, ID
William J. Archambeault, Goleta, CA
Victoria Arlak, New York, NY
David Arnold, San Francisco, CA
Daniel W. Arthur, Waynesboro, PA
Samuel L. Atmore, Waterville, ME
Jane Axelrod, Marblehead, MA
Joann Bally, Redondo Beach, CA
Jeff Barrow, Columbus, OH
Rex Bavousett, Kerrville, TX
Doug Beasley, Minneapolis, MN
Jim Benedict, Moose, WY
Tim Bennett, Upper Darby, PA
Greg Benson, Philadelphia, PA
David Todd Bintzler, Milwaukee, WI
Bobbie Blazy, Willoughby, OH
Betsey Bolton, Lowell, MA
Lisa C. Bord, Youngstown, OH
Elizabeth Ann Bowen, Tampa, FL
Kent Bowser, Columbus, OH
Charles Braendle, Phoenix, AZ
Jo Braund, Chesterland, OH
Stephen Bristow, Davis, CA
Arthur M. Brown, Tampa, FL
Leslie Brown, Springfield, MO
Annik Brunet, New York, NY
Stephen Burke, Corona, CA
Christopher Burnett, Albuquerque, NM
Steve Burns, Tacoma, WA
Cheryl Busbee, New York, NY
Kimberly Butler, Rego Park, NY
T. J. Button, Rocky Hill, CT
Michelle Byrne, West Carrollton, OH
John Cameron, San Francisco, CA
Ronald Caplain, Fall River, MA
Linda Carpenter, Winterpark, FL
Chuck Carver, New Hope, MN
Gary Cawood, Rustin, LA
H. R. Cellini, Albuquerque, NM
Stacy Cervantes, Commerce, TX
James Chotas, New York, NY
Larry Cincotta, Cooper City, FL
Roy J. Cirigliana, Dallas, TX
Sandra Russell Clark, New Orleans, LA
Brian Coats, Commerce, TX
Jim Colando, East Lansing, MI
Joe P. Cole, Ft. Worth, TX
Robert H. Cones, Brooklyn, NY
Karen Cooke, Chicago, IL
Michele Cousins, Indianapolis, IN
Ronald G. Crampton, Yelm, WA
David Crosby, Hickory, NC
Robert Csehak, Naugatuck, CT
Matthew Cullinane, Piedmont, CA
Curtis Culp, Salem, OR
Frank Curtis, Gainsville, FL
Gayle Curtis, Menlo Park, CA
M. P. Curtis, Columbus, OH
John A. Daly, San Francisco, CA
Brad Danielson, Columbus, OH
Bob Davis, Dayton, OH
Paul E. Davis, Simsbury, CT
Pati DeCesaro, Columbus, OH
Norb J. DeKerchove, Jr., Boise, ID
Francois Deschamps, New Paltz, NY
Rita Dibert, Mt. Baldy, CA
Sharon Dilworth, Minneapolis, MN
Kerry Dion, Reno, NV
Martha Dougherty, W. Alexander, PA
Nina Dudley, Cambridge, MA
Lin Eagle, Chicago, IL
Jane Eanes, Cartersville, VA
Barbara Edwards, Tallahassee, FL

Debbie Egan, Brooklyn, NY
Michael Elek, Munhall, PA
Jill Enfield, New York, NY
Steven L. Erholm, Anacortes, WA
Loret Falkner, Bloomington, IN
Rick Ferncase, Costa Mesa, CA
Abe Frajndlich, Cleveland, OH
Rick Farmer, E. Peoria, IL
William Faust, Charlottesville, VA
Richard Felber, New York, NY
D. R. Fiery, Earlyville, VA
R. L. Finkelstein, Catonville, MD
Flash Back, Columbus, OH
Rubie Floyd, Neptune Beach, FL
Judy Flynn, New Bedford, MA
Kennita Freed, Ann Arbor, MI
SFC Robert C. Freed, APO, NY
Susan Friedewald, San Francisco, CA
Nick Fuhr, Santa Barbara, CA
Diane Gager, Jacksonville, FL
Christopher Gallagher, Columbus, OH
Myrian Galler, Los Angeles, CA
O. Joseph Galli, Cincinnati, OH
Byron D. Gallup, Stamford, CT
Judy Leslie Gannon, Van Nuys, Ca
Gary Gantert, Jackson, MA
Lisa Garcia, New York, NY
Penny Gentieu, New York, NY
Monte H. Gerlach, Ithaca, NY
Robin Germany, Denton, TX
Neville Godfrey, Arcata, CA
Steve Goff, Lakewood, OH
Judith Goodman, Washington, DC
Paul J. Gordon, Upper Saddle River, NJ
John Gould, Denver, CO
Thom Greco, Milton, NY
Mike Greer, Lyons, IL
J. S. Greider, Sunnyvale, CA
John Griffin, Queens Village, NY
Danny Guthrie, Ithaca, NY
Florence Hales, Chicago, IL
Agnes Halpern, Oakland, CA
Genevieve Harm, Pittsburgh, PA
Marie Harward, Kensington, MD
Bonnie Hawthorne, Pacific Grove, CA
Timothy R. Hearsum, Reddings, CA
Craigie Huston Hemenway, Oswego, NY
William Herman, Evanston, IL
Lela Hersh, Urbana, IL
Ann Ginsburgh Hofkin, Long Lake, MN
Jon Hoggatt, Commerce, TX
Joe Holm, Lincoln, NE
John Hooper, Jr., San Antonio, TX
John Horna, Commerce, TX
Jean Hutchison, Tallahassee, FL
Richard Hutter, San Francisco, CA
Scot Indermuehle, Minneapolis, MN
Infrared Corp., Boise, ID
Peter Iverson, Tallahassee, FL
Sidney Jason, Los Angeles, CA
William Jenisch, Hancock, NH
Karen L. Jenson, Chicago, IL
Ann Johnson, Chicago, IL
Richard Johnson, Rustin, LA
Marguerite Rule Johnstone, New York, NY
Randy Juster, Bogata, NJ
Bruce Katsiff, Lumberville, PA
Robert Kendrick, Commerce, TX
Susan Kenney, Charlottesville, VA
Mark L. Ketter, Elkgrove, IL
Clarice Kjerulff, Weehawken, NJ
Jane Koegel, San Francisco, CA
William Koplitz, Sarasota, FL
Michael Koranda, S. Euclid, OH
Bobby Noel Kramer, Highland Park, IL
Mark Krastof, Chicago, IL
Helmut Kukowski, Pontiac, MI
Anne Kurutz, Rockport, ME
John Kustron, Columbus, OH
Jeff Kyler, Tampa, FL
Francesca Lacagnina, Seattle, WA
Carol Lachata, Pasadena, CA
Robbin Ladd, Redondo Beach, CA
Pamela J. Landau, Hoboken, NJ

Seraphina Landgrebe, Aptos, CA
Michael LaNoue, Whittier, CA
Marc A. Lansky, Fort Wayne, IN
Kathleen Larkin, Lavallette, NJ
Wendy Larkin, Phoenix, AZ
A. S. Laundon, Waltham, MA
Craig Law, Logan, UT
Tom Layman, Champaign, IL
Lisa Lazarus, Neptune Beach, FL
Gail LeBoff, New York, NY
Dennis Letbetter, San Francisco, CA
Judy Leslie, Van Nuys, CA
Shiu Leung, Nashville, TN
Michael Leuzzi, Paramus, NJ
Earl Levels, Richmond, CA
Richard J. Levy, Los Angeles, CA
Jennifer Lewis, Pasadena, CA
Brenda L. Lewison, Hudson, OH
John Lingerfelt, Salem, OR
Corrine E. Loomis, Salem, OR
Joseph John Lowry, St. Louis, MO
Richard Luken, Brookline, MA
Mike Lukowski, Omaha, NE
Giorgio Majno, Carbondale, IL
Gwen Manfrin, Walnut Creek, CT
Jed Manwaring, Corvallis, OR
Ken Marchionno, Dover, DE
Alissa Margulies, New York, NY
Charlie Martin, Los Angeles, CA
Frank Martin, Warren, MI
Randy Mayor, New York, NY
Dan McCormack, Accord, NY
Dianne McMillen, Captain Hook, HI
Sandra Merritt, Washington, DC
Scott Metcalf, Commerce, TX
V. E. Methvin, Tampa, FL
Kenn Michael, Allentown, PA
Elizabeth Miller, Garden City, NY
Jerome Miller, Detroit, MI
Marvin E. Miller, Palm Bay, FL
Philip Miller, Carmel, CA
Wesley Miller, San Francisco, CA
Edward Millet, Jr., Jackson, MS
Adam Misztal, Cleveland, OH
Megan Mitchell, Kealakekua, HI
James Modafferi, Brooklyn, NY
Stephen Molnar, Jr., San Francisco, CA
Richard Moore, Miami, FL
Susan J. Moore, Daytona Beach, FL
Christine Moran, San Francisco, CA
Gavin Morrison, Chicago, IL
Karen M. Muth, Conshohocken, PA
Jeffery Newbury, San Francisco, CA
Terry M. Newton, Livermore, CA
Diane M. Nicastro, Rochester, NY
R. M. Nottke, Schenectady, NY
Patty O'Brien, Buffalo, NY
David O'Connor, New York, NY
Dale O'Dell, Houston, TX
J. Keith Ostertag, Louisville, KY
Lori S. Owen, Kalamazoo, MI
Alfred S. Pagano, San Diego, CA
Richard R. Palie, Norwood, MA
David Paternite, Akron, OH
Stephen Paternite, Akron, OH
Louis Paulson, El Sobrante, CA
John Payne, Chicago, IL
Gregory J. Pels, Charlottesville, VA
Boyd Pennock, E. Peoria, IL
Lori Peterson, Urbana, IL
Margaret Peterson, Elmhurst, IL
Thomas Pfeffer, New York, NY
Mike A. Pfeifer, Fargo, ND
Debra Sue Phillips, Howe, IN
Tom Piccininni, Bronx, NY
Ellen Pinkham, Chicago, IL
Kay S. Pitkin, Minneapolis, MN
Tom Ploch, Santa Barbara, CA
Nicki Prell, Des Plaines, IL
Judith Preston, Eugene, OR
Cleofus Price, Brooklyn, NY
George Post, Berkeley, CA
Tod Quirk, Randolph, VT
Penny Rakoff, Akron, OH
April Rapier, Houston, TX

Kristin Mueller Read, Ann Arbor, MI
Elizabeth A. Reed, Bloomington, IN
Emmy Reese, Pacific Grove, CA
Patrick Renschen, Murphysboro, IL
Abe Rezny, New York, NY
Capt. Michael E. Richards, Los Angeles, CA
Ronald J. Rigge, Pleasanton, CA
Stephen Ritchie, Cantrall, IL
Allen V. Roberson, Houston, TX
Nancy P. Roberts, Milton, MA
Maura Robinson, Richmond, UT
Annie Rogers, Honolulu, HI
Hans Ropers, North Canton, OH
Ross K. Rorvig, Rothsay, MN
Terry Rozo, New York, NY
Debra Rueb, Houston, TX
Steve Ryan, Lincoln, NE
David Sachter, Rochester, NY
Hans Sander, Jersey City, NJ
H. J. Saunders, Vermilion, OH
Greg Savage, Butler, PA
Philip Schertz, Commerce, Tx
Erv Schroeder, Bloomington, IL
Lisa Schwabauer, Hubbard, OR
Robin A. Schwartz, Brooklyn, NY
Steven Schwartzman, Austin, TX
Bob Sebree, College Station, TX
Stan Shaw, Newark, OH
Susan Shea, Boston, MA
Michael Sheets, Toledo, OH
Randi Shepard, Highland Park, IL
Thomas H. Shuler, Jr., Washington, DC
Terry E. Simmons, Wellsville, NY
Claudia Simms, Miami Springs, FL
Ihor Slabicky, Newport, RI
Mark H. Sloan, Richmond, VA
Gerta P. Sorensen, DeKalb, IL
Kenneth Spector, Missoula, MT
Stephen Spera, Philadelphia, PA
Kathleen St. Peter, Hayward, CA
Susan Stein, Brighton, MA
Greg Wahl-Stephens, West Linn, OR
Carol Stoops, Minneapolis, MN
Marsha Fall Stuart, Auberndale, MA
Linda L. Sutton, Carlisle, PA
Lynn Swigart, Peoria, IL
Jay K. Switzer, Neoga, IL
Mary Z. Szot, Vacaville, CA
Hide Takei, Atlanta, GA
Mark P. Taylor, Jensen Beach, FL
Brad Temkin, Skokie, IL
Craig Tevis, Rexford, KS
Frank E. Thomson, Columbia, SC
Jim Toms, Arcata, CA
Michael Trojansky, Philadelphia, PA
Brian C. Trombley, Orangevale, CA
James Uhl, Kingston, PA
Thomas Upton, Palo Alto, CA
Susan Vadasz, Riverview, FL
Bruce Van Meter, Trinidad, CA
Mike Vazquez, Claremont, CA
Nicole Verhagen, Henderson, NV
Jean S. Vitatoe, Richmond, IN
Edward Waddell, Charlotte, NC
Mary Walsh-Lahti, Volcano, HI
Dr. Robert C. Waltz, Lyndhurst, OH
Pat Ward, San Antonio, TX
R. D. Weidenbusch, Columbus, OH
Mark C. Weidling, Glenview, IL
Norman Weinstein, Boise, ID
Harris Welles, Lawrenceville, GA
Paula West, New York, NY
Nancy H. Whitehead, Covington, KY
Mia Widell, Laguna Beach, CA
Dion L. Wilson, Peoria, IL
Merry Moor Winnett, Greensboro, NC
Robyn Wishna, Providence, RI
John Wong, San Francisco, CA
Sheryl L. Wood, Stuart, FL
Mike Woodside, Rochester, NY
Bob Wright, Jensen Beach, FL
Richard Wunsch, Wausau, WI
Debra Zare, Libertyville, IL